Collins

THE SUN

T0364516

Dr Ryan French

Published by Collins
An imprint of HarperCollins Publishers
Westerhill Road, Bishopbriggs, Glasgow G64 2QT
www.harpercollins.co.uk

HarperCollins Publishers
Macken House, 39/40 Mayor Street Upper,
Dublin 1, D01 C9W8, Ireland

In association with
Royal Museums Greenwich, the group name for the National Maritime Museum,
Royal Observatory Greenwich, the Queen's House and *Cutty Sark*
www.rmg.co.uk

© HarperCollins Publishers 2023
Text © Dr Ryan French
Cover photograph © NASA Solar Dynamics Observatory/Ryan French
Images and illustrations: See Acknowledgments page 106

Collins ® is a registered trademark of HarperCollins Publishers Ltd

All rights reserved. No part of this publication may be reproduced, stored in a retrieval system,
or transmitted, in any form or by any means, electronic, mechanical, photocopying, recording
or otherwise without the prior written permission of the Publisher and copyright owners.

The contents of this publication are believed correct at the time of printing.
Nevertheless the Publisher can accept no responsibility for errors or omissions,
changes in the detail given or for any expense or loss thereby caused.

HarperCollins does not warrant that any website mentioned in this title will be provided uninterrupted,
that any website will be error free, that defects will be corrected, or that the website or the server that
makes it available are free of viruses or bugs. For full terms and conditions please refer to the site
terms provided on the website.

A catalogue record for this book is available from the British Library

ISBN 978-0-00-858023-0

10 9 8 7 6 5 4 3 2

Printed in India by Replika Press Pvt. Ltd.

If you would like to comment on any aspect of this book, please contact us at the above address or online.
e-mail: collinsmaps@harpercollins.co.uk

facebook.com/CollinsAstronomy
@CollinsAstro

This book contains FSC™ certified paper and other controlled
sources to ensure responsible forest management.

For more information visit: www.harpercollins.co.uk/green

Contents

Warning

Before we begin, let us get one thing clear. Although the human eye is well adapted for many things, looking directly into the Sun is not one of them. And so, despite the alluring image on the front page of this book, I urge you never to look at the Sun's surface directly with the naked eye. The pages that follow are full of tales old and new, of how our scientific predecessors observed the Sun, how scientists observe the Sun today, and how you can observe it safely for yourself. But make no mistake, looking directly at the Sun with incorrect equipment may result in permanently altered colour perception, blind spots, distorted vision, and so on. But do not fear: the Sun is a beautiful object and one worthy of appreciation, and there are many ways to do so safely – stay tuned.

Introduction

As it sits above our heads on a sunny day, it is easy for us to become desensitised to the Sun. It rises, sets, and appears to do nothing in between, beyond providing us with light and warmth. This, in itself, is worthy of recognition (as our ancestors did centuries ago), but unbeknownst to the unaided human eye, the Sun is far more complex than this. Despite its seemingly unchanging appearance, the Sun is incredibly dynamic, with an atmosphere continuously churning and changing, evolving over timescales from seconds to beyond centuries. Some of these processes on the Sun, such as *solar flares* and *coronal mass ejections*, can affect our lives here on Earth. These solar processes, although far away, can damage our satellites, block our communication networks, and even shut down our power grids. It is, therefore, worthwhile, for us as a species, to understand our local star and its connection with Earth.

We owe the Sun a lot. If it had formed a little bit hotter or slightly cooler, you would certainly not be reading this book today. The Sun is about 109 Earth widths in size from edge to edge, about 1.3 million Earths in volume (109 x 109 x 109), and the Earth orbits the Sun at a distance of about 107 Sun widths (150 million km). Despite these large numbers and its significance to us, the Sun is a surprisingly unremarkable star, with an average size, and one of around 250 billion stars within the Milky Way Galaxy alone. But unlike these other stars, the Sun has one major advantage – we can observe it. Observations of other stars in the galaxy are mostly limited to a one-dimensional point of light, but we observe the Sun in high resolution many times each minute, with telescopes sensitive to the entire spectrum of light from radio waves to gamma rays. Scientists researching the Sun are, therefore, opening a wider door to the universe, applying knowledge of our local star to all the stars beyond it.

The purpose of this book is to introduce you to the Sun as if it were your first time learning about it. We will explore the journey our scientific predecessors took in understanding the Sun and the role the Sun played in the lives of our distant ancestors long before them. By the end of this book, you will have a grasp of some of the physics behind physical processes on the Sun, from *nuclear fusion* in its core to eruptions on its surface, and learn how this information relates to distant stars, *exoplanets*, and beyond. Finally, we will get you involved, with resources on how you can access NASA and ESA observations of the Sun from the comfort of your own home, or observe the Sun safely from your garden. Whether you want to observe on an ordinary sunny day, or during a total solar eclipse, this book has you covered.

Welcome to the Sun.

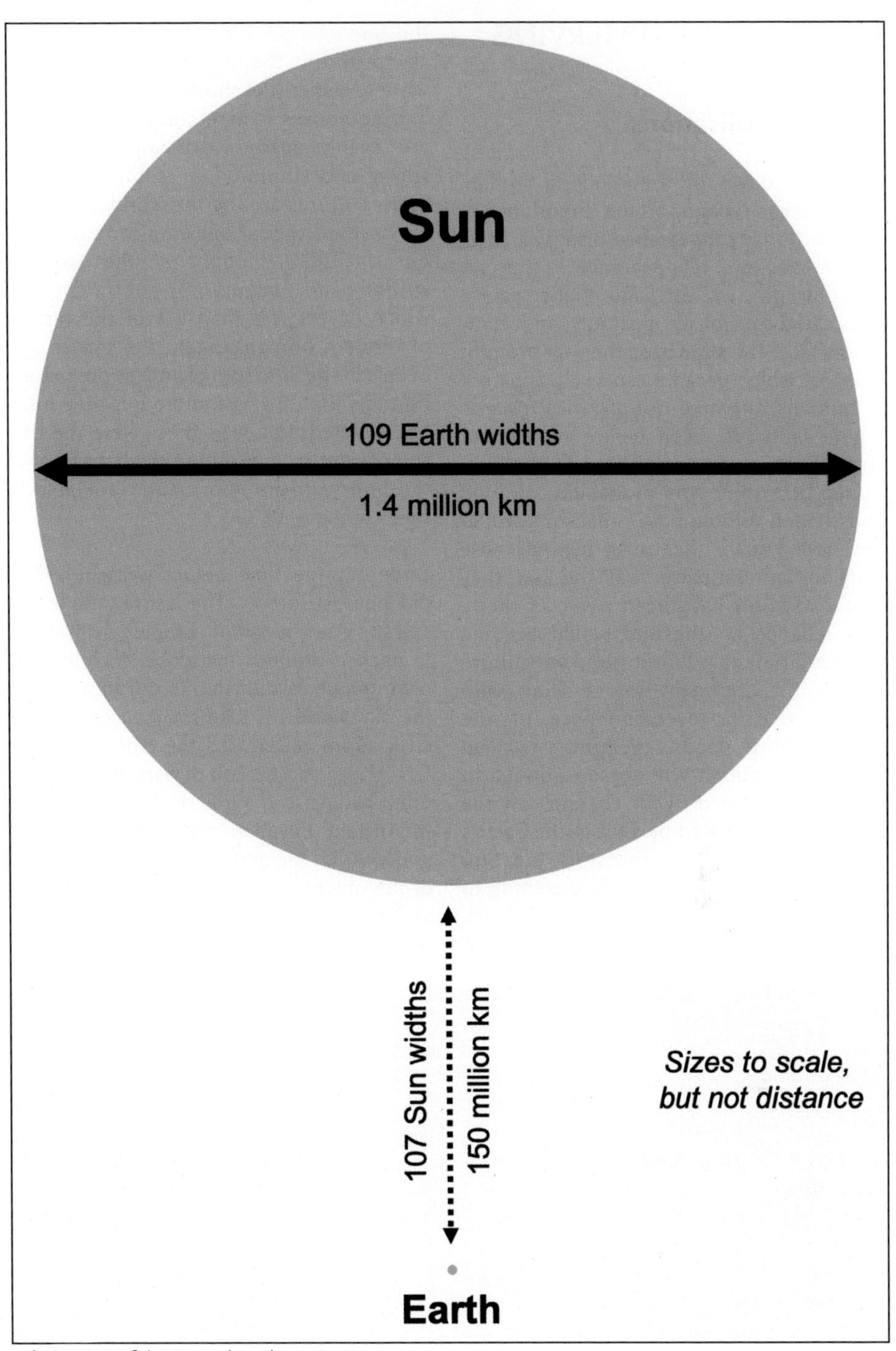

Relative sizes of the Sun and Earth.

1: HISTORY OF OBSERVING THE SUN

Ancient Civilisations

Since the dawn of the modern human, mankind has recognised the importance of the Sun. Its daily appearance, disappearance, and subsequent reappearance in the sky brought light, warmth, and relative safety. The variation in its position and daily longevity in the sky across the year brought seasons, which became especially apparent as humans migrated out of Africa towards higher latitudes. Even before the age of farming, the seasons, brought by the Sun, would determine the availability of food and shelter. Although our ancestors would not have known the cause behind these daily and annual patterns of the Sun, they would certainly have been aware of them. This reliance on the Sun would become more prominent as farming and agriculture became the dominant way of living, with the absence or overabundance of the Sun bringing floods, drought, or failed harvests. Although the seasons on Earth have nothing to do with changes on the Sun (they are created by a tilt in the Earth's rotation axis), to our ancestors the Sun would have been the most obvious thing to assign responsibility to for these variations. Even today, the summer solstice (when the Sun reaches the highest point in the sky, providing the longest day of the year), winter solstice (the shortest day of the year), and equinoxes (where sunrise and sunset are evenly spaced, ocurring once each spring and autumn), are still important in many cultures around the world. However, we have no way of knowing for sure what our ancestors thought of the Sun, as written history begins only 5,000 years ago, which covers less than 2% of the history of modern humans. Sadly, the stories and beliefs of the Sun from countless prehistoric humans are now lost, although they must have existed. In a way, these were the first solar scientists, explaining the behaviour of the Sun with the information available to them at the time.

Evidently, the time before written history did not last forever. The last few millennia saw the rise and fall of complex civilisations in pockets arounds the globe, each with its own unique relationship to our local star in the sky above us. Like their ancestors, the civilisations recognised the critical role the Sun played as a beacon of warmth and light, often hailing it as a god or deity as a result. In Ancient Egyptian mythology, the chief god was Ra, the god of the Sun and creator of the universe. Ra has the head of a falcon and carries the Sun on his head within his headdress. It was believed that the motion

Cave painting of the Sun in Sedona, Arizona.

Ancient Egyptian god of the Sun, Ra.

of the Sun in the sky was Ra sailing across the sky in a boat.

In early ancient Greece, the story was similar. Instead of a boat, the Sun was believed to be pulled across the sky by a horse-drawn chariot, ridden by either the Titan Helios or the god Apollo (depending on the exact period and location within ancient Greece). Helios was an earlier figure in Greek mythology and not assigned a particularly great deal of importance. The word *helio* today derives from the Titan Helios and is used to define something relating to the Sun (e.g. heliocentric, heliosphere, words that you will see later in this book). Apollo was more of a significant figure in ancient Greece. As a son of Zeus, Apollo was the god of many things, including archery, dance, poetry, light, and, of course, the Sun. You may have heard the word *Apollo* in relation to space before, as *Apollo* was the name of the NASA spaceflight missions that landed 12 men on the Moon between 1969 and 1972. The name choice for the mission was bizarre, as Apollo was the god of the Sun, not the Moon. NASA's next crewed mission to the Moon, *Artemis*, scheduled for later this decade, has a far more sensible name, as in Greek mythology Artemis was the twin sister of Apollo and the goddess of... you guessed it, the Moon. The ancient Greeks eventually came to learn a lot about the Sun. By 600 BC, they had determined it to be a hot ball of gas and even studied the orbits of the Sun and Moon well enough to predict solar eclipses (although they did still believe the Sun orbited the Earth).

There is a large overlap between ancient Greek mythology and that of ancient Rome. There are many equivalents of ancient Greek gods in Roman mythology, such as between Zeus and Jupiter, Poseidon and Neptune, and so on. Much like the Greek gods Apollo and Artemis, the Romans also believed in a sibling pair of a god and goddess of the Sun and Moon. For the Romans, these were Sol and Luna. Much like the Greek word *helios*, the latin words *sol* and *luna* still exist in the English language to describe the Sun and Moon (e.g. solar, lunar).

Helios and his chariot displayed on a vase.

Huitzilopochtli, the Aztec god of the Sun and war.

The existence of a Sun deity was not just a European and northern African phenomenon either. Sun gods were worshipped all over the world, from ancient China to the Incas and pre-Islamic Arabia. In northern Mexico, the people of the ancient Aztec Empire also worshipped a Sun god – Huitzilopochtli. For the Aztecs, Huitzilopochtli was the deity of war, the Sun, and human sacrifice. He wielded a flaming serpent as a weapon; hence his association with the fiery Sun in the sky. They believed human sacrifice to Huitzilopochtli was necessary in order for him to rise the next day and protect them from the infinite night. Thankfully, there is no human sacrifice involved with studying the Sun today.

Humanity's collective knowledge of the Sun has grown significantly over the past few hundred years, and solar physicists (scientists researching the Sun) today have access to a wide range of telescopes, spacecraft, and computing power, the likes of which our scientific ancestors could never have imagined. When Galileo pioneered the use of telescopes for astronomy in 1609, did he ever consider that one day telescopes would soar through the solar system, untethered from planet Earth?

A Monk's Tale

Imagine, if you will, that you're a monk in the English town of Worcester in 1128 AD (perhaps you are already a monk, in which case you are already halfway there). You wake up, doing whatever it is English monks typically did on such a morning, and head outside, as you would on any other day. But today, something is different. It is a misty morning, and the Sun has recently risen. The combination of the Sun sitting close to the horizon and the thick, misty air means that it is possible to look directly at the Sun for a few brief moments. (Please note: even under these conditions, you should **never** look directly at the Sun). In the fleeting moments that you, the 12th-century monk, catch a glimpse of the Sun, you notice something peculiar. Instead of the uniform circular disc you've become accustomed to seeing your entire life, you see two large circles on the Sun, sat either side of the Sun's equator. The image on the next page shows the drawing made of the Sun on this day in 1128. If you imagine you are the monk who saw this, with almost no prior knowledge of the Sun, what would you think of this? How would you explain what you had witnessed? At this point in history, Europeans believed the Earth to be the centre of the universe, with the Sun, Moon, and stars orbiting around us – constant, perfect, and unchanging. If this were you, might you have thought you were witnessing a hole appearing in the Sun? Perhaps a tunnel? What lay within that tunnel? Perhaps a dragon, or a demon maybe? What would you have thought?

The drawing was made by John of Worcester within the Worcester Chronicles, an important text of English history documenting the period between 734 and 1140 AD. John was the last of many authors of the chronicles. We will never know what John truly thought of such a sight, as to

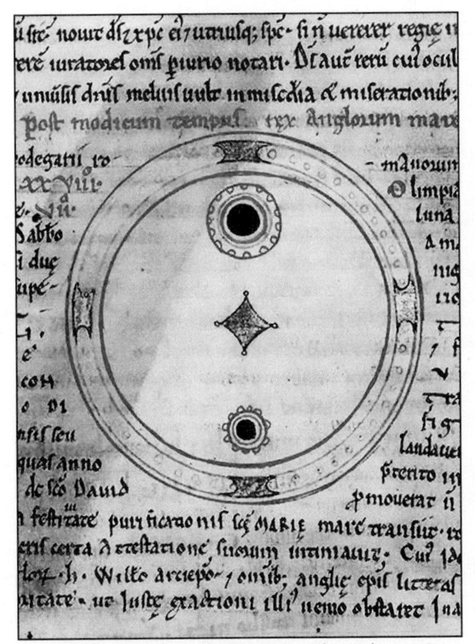

Sunspot diagram from the Worcester Chronicles, 1128 AD.

over-speculate on imperfections in the sky above could have been considered heresy at the time. Whatever he thought, though, he certainly thought it important. Within the hundreds of pages of text written over 406 years, the Worcester Chronicles contain only five images – three depicting dreams of King Henry I, one of the crucifixion of Jesus Christ, and one of the Sun on that specific day. John also describes a solar eclipse viewed two years later in 1130, but apparently considered it relatively unnoteworthy, as he offered no illustration of the cosmic event.

John of Worcester's drawing of these strange, dark features on the Sun turned out to be important after all, as he had just recorded the first ever observations of *sunspots*. It is possible that such sightings had been made before, but no known record of them exists. Observations of sunspots would not be recorded again for another five centuries, until Galileo Galilei and the invention of the telescope.

Squabbling Astronomers and the Era of the Telescope

In 1609–1610, the world of astronomy changed forever. With the use of a refracting lens, Italian astronomer Galileo Galilei created a telescope and pointed it to the sky for the first time. (Who exactly to give credit to for inventing the telescope is not an easy question, but Galileo was certainly the first to use it for astronomy). In the coming years, Galileo was the first person ever in human history to view the rings of Saturn, the four largest moons of Jupiter (which now bear his name), and mountains on the Moon. It also wasn't long before he pointed his new invention at – you guessed it – the Sun. Originally limiting his telescope observations of the Sun to early sunrise and late sunset to restrict damage to his eyes, he eventually switched to the much safer projection method of observing the Sun (which we will explore later in this book).

Upon looking at the Sun, he immediately noticed clusters of small black spots – sunspots. Galileo made systematic observations of the Sun, sketching the configuration of sunspots as they evolved and moved across the Sun's surface. He noticed that, like the Earth, the Sun

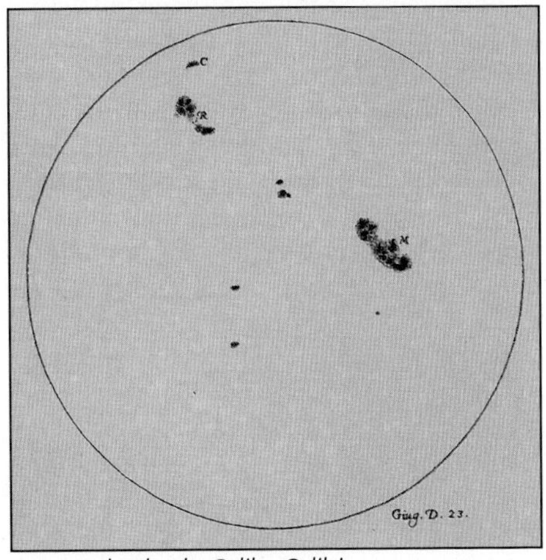

Sunspot sketches by Galileo Galilei.

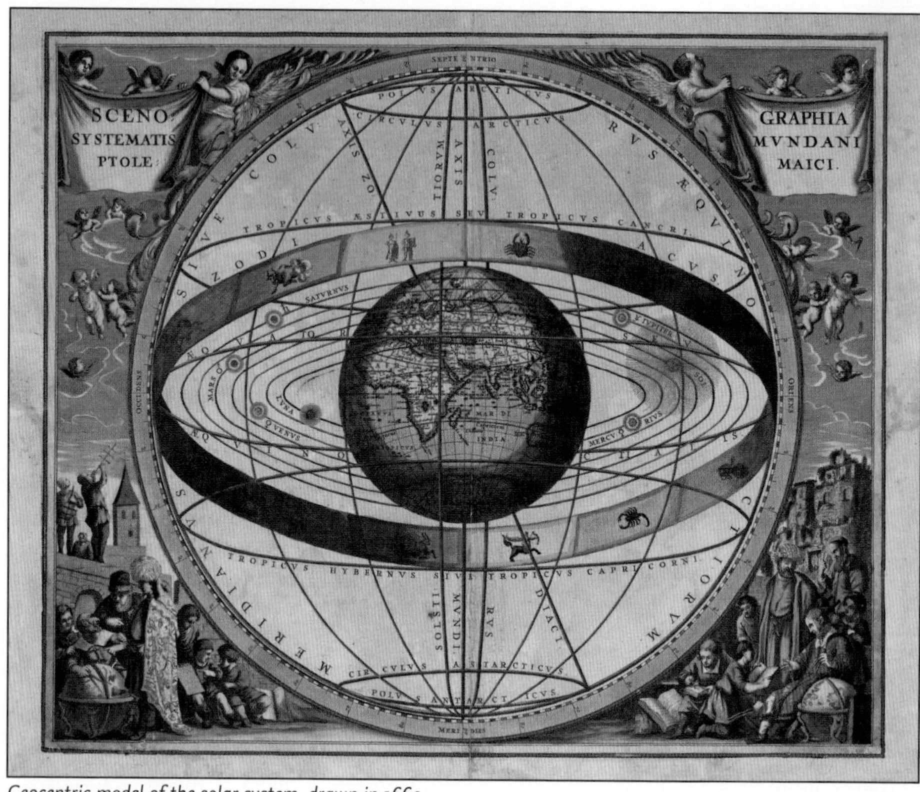

Geocentric model of the solar system, drawn in 1660.

appeared to be rotating, as sunspots and other features appeared over the left edge of the Sun, moved across the Sun's face, and disappeared over the right edge, all within a couple of weeks or so. He was also able to identify sunspot structure, noticing darker inner regions of the sunspots surrounded by slightly less dark areas (which were still darker than the surrounding bright Sun). Galileo also recorded that sunspots would evolve from day to day, sometimes completely appearing or disappearing over a few days. An example of a sunspot map sketched by Galileo is shown in the image on the previous page.

Galileo's pursuit of knowledge of the Sun and night sky unfortunately landed him in a lot of trouble. The more he saw, the more he became convinced of the *heliocentric* model of the solar system – the idea that the Sun is at the centre of the solar system, not the Earth. Such a concept was fresh and unwelcome news within Europe (and particularly in Galileo's home country of Italy) at the time. The concept was first suggested by the Polish mathematician and astronomer Nicolaus Copernicus a few decades earlier, in 1543. Galileo eventually stood trial for heresy in 1633 for his heliocentric suggestions, in which he entered a plea. He was spared his life in exchange for house arrest, where he spent the rest of his life continuing work from his home until his death on Christmas Day in 1642, at the age of 77.

Long before his trial and house arrest, the Sun was causing Galileo other troubles. Despite Galileo's frequent and

Heliocentric model of the solar system, drawn in 1661.

highly detailed drawings of sunspots, neither he, nor the rest of the astronomy community, had any clear idea of what they were looking at. Unlike moons orbiting Jupiter or mountains on the Moon, the concept of dark spots on the Sun was completely abstract and unknown. And so, in a textbook example of how the scientific method should work, astronomers began the process of attempting to explain the origins of these bizarre sunspot features.

One of the first scientists to take a punt at solving the mystery of sunspots was the German astronomer Christoph Scheiner. Scheiner postulated that sunspots were not related to the Sun at all but were instead 'small planets closely orbiting the Sun'. This hypothesis was quickly debunked by Galileo, who showed that sunspots rotated with the Sun, not independently of it. The inner planets of the solar system, Mercury and Venus, do actually pass the Sun in this way, but look very different from sunspots. Although Scheiner was wrong in this instance, the concept itself was certainly ahead of its time. For today, centuries later, exoplanets (planets orbiting stars other than our own Sun) are discovered in this manner, by observing a periodic drop in a star's brightness as planets pass in front of the star from our perspective. But alas, Scheiner lived far too early for exoplanet science. And so, as is the case with the scientific method, with the addition of new evidence contradicting the standing theory, the theory must either evolve to explain the contradiction or be scrapped. In this case, it was the latter.

With the discovery that sunspots rotate with the Sun, Galileo suggested that they were 'perhaps cloud-like structures in the solar atmosphere'. Looking at sketches of sunspots at the time, you can see how this idea may have formed. The lighter boundaries of the sunspots do appear somewhat wispy in texture, so it was not a giant leap to equate them to equivalently wispy features in the sky here on Earth – clouds. Christoph Scheiner, likely still remembering Galileo's rebuttal of his earlier theory, bounced back again with a different theory to counter Galileo's. Scheiner argued that instead of cloud-like features in the Sun's atmosphere, sunspots behaved more like 'dense objects embedded in the Sun's luminous atmosphere'. That is, they were more analogous to islands sitting in the sea than clouds in the sky. In the end, neither Galileo nor Scheiner were correct in their theories, but neither of them would live to know it – as no significant progress was made in the understanding of sunspots for another 200 years.

Enter William Herschel

William Herschel was born in Germany in 1738 and lived a long life full of achievement. One of Herschel's biggest accomplishments came in 1785, when he was awarded £4,000 by the king of Great Britain to build an unimaginable scientific instrument, in which he succeeded. He built a 12-metre-long telescope, with a 1.26 metre primary mirror at the bottom. This design of telescope, which used a mirror instead of the lens in Galileo's original design, was first introduced by Isaac Newton over 100 years earlier, but this was the largest implementation of it to date (by far). In fact, Herschel's '40-ft telescope', as it's known, was considered the largest scientific instrument ever created at the time. This claim really depends on what you classify as a scientific instrument, as some may argue that structures like Stonehenge, or even the pyramids, which were built to align with the stars, could be considered as scientific instruments in some way. Nonetheless,

Herschel's '40-ft telescope'.

Herschel's creation was glorious and opened up the sky to new discoveries. Even by today's standards, Herschel's '40-ft telescope' would have been considered a decent telescope. His most famous discovery (with an earlier telescope), was of the planet Uranus, the seventh planet from the Sun and the seventh planet known to us at the time. Neptune, the eighth planet of the solar system, was not discovered until 1846 – 24 years after Herschel's death. William Herschel also discovered two moons of Saturn: Enceladus, a moon famous for having a global water ocean under an icy surface, and Mimas, a moon famous for looking a bit like the Death Star from *Star Wars*. Herschel and his work were well respected within the astronomy community and Herschel was made the first ever president of the Royal Astronomical Society in 1820, shortly before his death two years later, at the age of 83.

Like many other night-time astronomers at the time, William Herschel also dabbled in solar physics, although his solar work was not as well respected in the field as his other work. In efforts to research the correlation between the Sun and the Earth's climate, he collected daily sunspot records for 40 years between 1779 and 1818. His methodology was certainly unique at the time, comparing trends in sunspot numbers with the price of wheat. His idea was that the price of wheat would vary with the annual supply of wheat, which would be related to temperatures in Europe. Finding a connection between sunspot numbers and the wheat price would provide evidence for cycles on the Sun affecting the Earth's climate. William Herschel found no such correlation in his data. The relationship between the Sun and Earth's climate is a complex and multifaceted one, and an ongoing area of research today. We do know that modern global warming is undoubtedly a result of human activity and not related to solar activity (which has been decreasing steadily on average since the 1980s). During his work collecting sunspot observations, Herschel did, like Galileo and Scheiner before him, hypothesise on the true nature of sunspots. He suggested that sunspots were 'Openings in the Sun's luminous atmosphere, allowing a view of the underlying, cooler surface of the Sun'. If you know anything about sunspots, which you will by the end of this book, you may think that Herschel's hypothesis was fairly accurate, far closer to uncovering the true nature of sunspots than Galileo, Scheiner or the 12th-century monks ever achieved (although Herschel had access to greater tools and pre-existing knowledge). Except, unfortunately, William Herschel's description of the nature of sunspots

Mimas and Enceladus, two moons of Saturn discovered by William Herschel.

did not end there. The full, unabridged statement is: sunspots are 'Openings in the Sun's luminous atmosphere, allowing a view of the underlying, cooler surface of the Sun, which is likely inhabited'.

That's right. William Herschel, the discoverer of worlds and first ever president of the Royal Astronomical Society, believed that there were habitable regions on the surface of the Sun. This seems like a ridiculous suggestion to us today, and it may seem tempting to accredit this outlandish claim to the fact that we know far more about the Sun today than Hershel did in the early 1800s. But even for that period, the suggestion of possible life on the Sun was a wacky one.

I believe this story of William Herschel to be a good lesson. Not only within solar physics or astronomy, but in life. It's always worth remembering that everyone, even the most qualified, accomplished, and brilliant people, make mistakes, disastrous and embarrassing ones, from time to time.

Discovery of the Solar Cycle

By the middle of the 1800s, the puzzle of sunspots would get even stranger. Heinrich Schwabe, a German astronomer, was about to make an accidental discovery that would change solar physics forever. There were a lot of 'accidental discoveries' in this premodern era of physics, as available scientific instrumentation would accelerate beyond collective scientific knowledge at the time. There was a lot of catching up to do – observing strange things and attempting to explain them with science. Nowadays, research in modern solar physics is very different. We have advanced theories, computational models, and simulations available to us which offer plausible explanations to puzzles about the Sun (which we will explore later). But the instrumentation available to us is only now reaching the level of offering firm observational proof of these possible solutions. Heinrich Schwabe's accidental discovery came as he searched for an extra planet in the solar system with an orbit closer to the Sun than Mercury. A planet very close to the Sun would be undetectable in the night sky, and so Schwabe planned to look for the potential planet by waiting for it to pass in front of the Sun. The passing of a planet in front of the Sun, as we occasionally see Mercury and Venus do (for they are closer to the Sun than us), is called a planetary transit. Mercury and Venus transit the Sun relatively infrequently, a few times a century between them, but a planet far closer to the Sun would do so more often (we'll discuss Mercury and Venus transits later in the book).

Schwabe's plan was to observe the Sun every day from 1826 to 1843, scanning the sunspots to search for a planet amongst them. He was ultimately unsuccessful in this endeavour, as no such planet exists, but what he did find was equally exciting. As Heinrich Schwabe looked back through his 17 years of sunspot observations, searching for a nonexistent planet, he noticed something peculiar. It was known at the time that the quantity of sunspots on the Sun was not constant but changed with time. Schwabe noticed a pattern in the frequency of sunspots and found in his data that the number of sunspots followed a cycle. He suggested that this sunspot cycle had a period of about 10 years – meaning every 10 years the cycle would reach a maximum number of sunspots. Considering he had access to only 17 years of data, this estimation of a 10-year cycle was remarkably close. In reality, sunspot numbers follow a cycle with an average period of 11 years. In the century that followed, many other scientists would solidify and push further our understanding of the solar cycle. Unfortunately for Schwabe, it is these scientists who are remembered today for their solar cycle research.

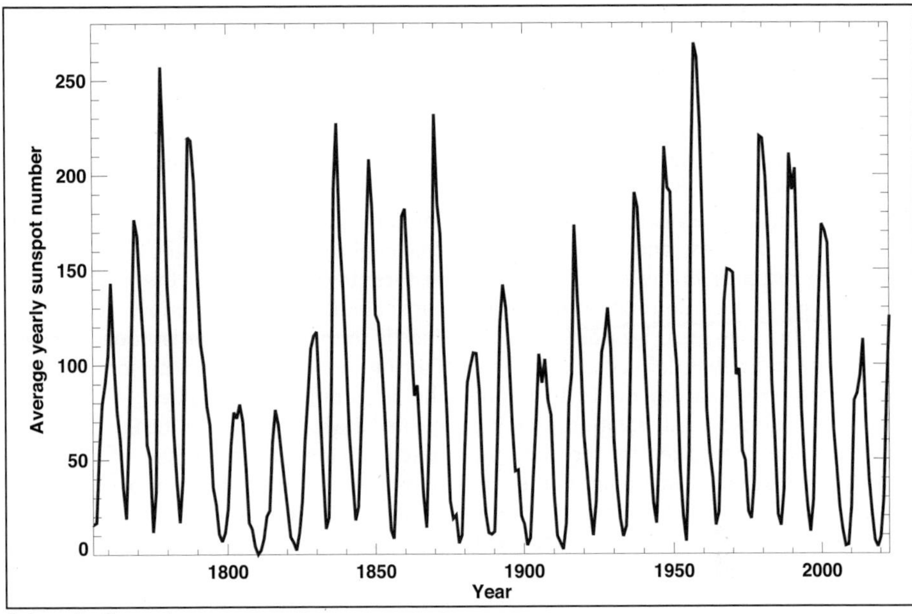

Solar cycle of sunspots, 1755 – 2023

After hearing of Schwabe's discovery of a cycle of sunspot activity, the Swiss astronomer Johann Rudolf Wolf took things further. Using historical sunspot data dating back to 1755, he discovered an average solar cycle duration of 11.11 years. Wolf also established a naming convention for the solar cycles. Unable to find suitable sunspot diagrams predating 1755, he named the 1755–1766 solar cycle as 'Solar Cycle 1'. At the time of Schwabe's work, the Sun was in Solar Cycle 8, and in Solar Cycle 9 by Wolf's work in 1852. This is still the naming convention we use for solar cycles today. In 2019, Solar Cycle 24 ended, and we entered Solar Cycle 25. Wolf was unable to find reliable sunspot observations prior to 1755 (the start of Solar Cycle 1), simply because they did not exist. Before 1755, the Sun was deep in a period of unusually low sunspot activity, and sunspots were not a regular enough occurrence to detect any remnant of an 11-year cycle. It is believed that if this period of low solar activity had not happened, previous astronomers, such as William Hershel, would have discovered the solar cycle much earlier.

A Part Time Astronomer and Brewery Owner

Richard Carrington was a part-time astronomer and part-time brewery owner, a great job combination even by today's standards. During the 1850s, Carrington also observed the Sun daily, just as Heinrich Schwabe had in the decades before. However, instead of doing so to search for a mystery planet, he was attempting to determine the rotation period (time of one full rotation) of the Sun. This task is not as simple as it sounds. The Earth, like Mars, Venus, and Mercury, has a solid and rigid surface.. Although the Earth's crust does evolve due to the molten mantle under the crust, it does so on far slower timescales than the planet's rotation. It is, therefore, easy to calculate the rotation period of Earth by picking an object in the sky, starting the timer, and stopping the timer when the object returns to the same east-west location (due to the Earth's tilt, the object will have drifted slightly north or south during this time). You can do this experiment anywhere on Earth with the same success. However, the answer you get will depend on which object in the sky you choose. If you choose any star, it will, to the nearest second, take 23 hours, 56 minutes, and 4 seconds for the star to reappear in the same east-west location. This is the true rotation period of the Earth. This may seem bizarre to you, as surely a day on Earth is defined as 24 hours? This is true, except a day on Earth is not defined by a single Earth rotation of 360°, but instead the time taken for the Sun, not the stars, to reach the same point in the sky. Therefore, in the 23 hours, 56 minutes, and 4 seconds it takes for the Earth to complete a single rotation, the Earth has moved farther in its orbit around the Sun. The Earth then needs to rotate for another 3 minutes and 56 seconds in order for the Sun to move to the same (east-west) location in the sky. The position of stars in the sky, therefore, drifts by 3 minutes and 56 seconds every night, which explains why the constellations of the summer night sky are different from those in the winter night sky (like the constellation Orion, which is only visible from October to April). Multiplying this 3 minute, 56 second time lag by the amount of Earth days in a full orbit (365.25), you'll get 24 hours, which means the stars will always be in the same position on a given date each year.

Calculating the rotation of other solid planets is also easy. Simply pick a recognisable feature on the surface of the planet and wait for it to rotate round to the same position. This will give you the relative rotation period between the planet and the Earth, and by considering how both planets have moved within that time, you can calculate the true rotation period. For planets without a solid surface, calculating the rotation is more difficult, as the rotation speed need not be the same at different depths beneath the surface, or even at different latitudes across the surface. This is called differential rotation and makes it difficult to track a specific surface feature over a full rotation. The measurement of Saturn's rotation, for example, has high uncertainties to this day.

Measuring the rotation of the Sun is challenging. Sunspots provide a clear feature to track across the surface, but sunspots at the equator move at different speeds to those at higher latitudes. Richard Carrington was the first person to study the Sun's differential rotation in detail and created a system to define the Sun's rotation based on the rotation speed of the Sun's mid-latitudes, which have an orbital period of 27.3 days. Rotation at the equator and poles differs from this value by a few days, with a period of 24 days at the equator and over 32 days at the poles. Carrington named this mid-latitude rotation, which reflects something close to the average rotation period across the entire Sun, as the 'Carrington rotation'. The Carrington rotation is still standard for measuring solar rotation today.

The Carrington Event

In the year 1859, Richard Carrington discovered something that would change our view of the Sun forever. During his routine observations of sunspots for his research on differential rotation, Richard Carrington made a discovery that shook the scientific community. Although nobody knew exactly what Carrington had witnessed, scientists understood that the findings were significant. On Thursday, 1st September 1859, Richard Carrington was observing the Sun as a part of his daily sunspot observations when he noticed something new. Unlike anything he had seen on the Sun before, Carrington spotted bright features within a group of sunspots. These bright features were not flashes, but instead persisted for several minutes as their shape, size, and location slowly evolved within the sunspot. The image below shows Carrington's sketch of these bright features, which he labelled A–D within the sunspot. If Carrington had been alone during this observation, he might not have been believed. Astronomers had been watching sunspots for over 200 years by this time, and never before had anyone witnessed anything on the Sun changing in real time. Fortuitously, miles away, another English astronomer, Richard Hodgson, independently made the same observation. This was certainly an intriguing thing for Carrington and Hodgson to witness, but the true discovery was yet to come.

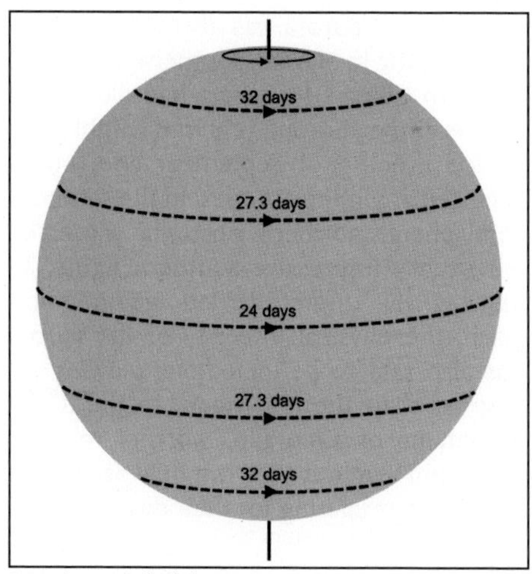

Rotation period of the Sun.

Less than 18 hours after Carrington and Hodgson's observations, the world witnessed the most intense aurora display it had ever seen. The aurora, more commonly known as the northern lights (*aurora borealis*) in the northern hemisphere and southern lights (*aurora australis*) in the southern hemisphere, has long been observed throughout human history, but typically confined only to high latitudes near the poles. But during this night of 1st September 1859, populations across large geographical regions, where the aurora had never been sighted previously, witnessed an impressive display beyond their comprehension. In North America, where it was night-time as the aurora began, the northern lights were reported as far south as Florida, Cuba, southern Mexico, and even Colombia. As night rolled across the Pacific Ocean and

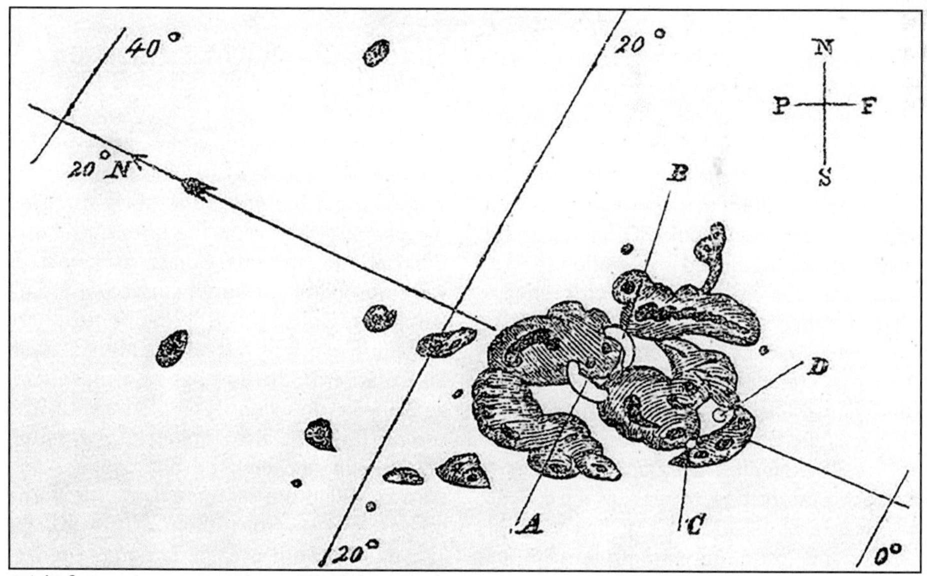

Bright features in a sunspot observed by Richard Carrington.

into Asia, the aurora was also witnessed for the first time in Hawaii, southern Japan, and southern China. The aurora had died down by the time Europe entered night-time on the evening of September 2nd, but it persisted at higher latitudes. In the southern hemisphere, northern Australia witnessed the equally impressive southern lights. As well as its unprecedented geographical reach, the physical intensity of the aurora was also said to be incredible, glowing far brighter than the full Moon. In the Rocky Mountains of Colorado, workers mistook the brightness out their window as sunrise and began preparing for their day.

Although Europeans missed the best of the northern lights display, they experienced equally bizarre happenings. In 1859, the height of technology was the telegraph machine – a big clunky machine that could send signals along telephone wires, to the recipient's telegraph instrument. This was before the invention of the telephone, so no human voices were communicating via these signals, only written messages. During the daytime of 2nd September, whilst the Americas were experiencing the aurora during their night-time, Europe began experiencing weird behaviour with their telegraph machines. Telegraph machines that were switched off and disconnected from any power source were continuing to send and receive messages. No electricity was powering them at all, and yet they were still functioning... spooky. Other telegraph machines that remained in operation sparked and gave operators small electric shocks. There are even some claims, including on the Wikipedia Page for this event, that telegraph machines literally burst into flames. However this claim is largely unsubstantiated, with little reliable historical evidence. I don't think you need exploding telegraph machines to make this interesting, though, as the concept of phantom telegraph signals is surely bizarre and intriguing enough (even if less visually exciting).

After hearing the stories of exceptional aurora and strange technological occurrences, Richard Carrington immediately put two and two together. Although he did not know any physical mechanism capable of doing so at the time, he was certain that the strange behaviour on Earth was related to the bright feature on the Sun he had recorded only a day earlier. Carrington had, for the first time, discovered that activity on the Sun's surface could influence us here on Earth. He had discovered *space weather*. The bizarre conditions that followed Carrington's observations, including the aurora and technology malfunctions, ended up bearing his name – it came to be known as *The Carrington Event*.

But how could this be? How could a bright light on the Sun 150 million km away create the aurora and telegraph malfunctions here on Earth? A little spoiler for later in this book – these were the result of a coronal mass ejection and *geomagnetic storm*. Unbeknownst to Carrington and other astronomers at the time, the answer actually lay hidden in eclipse observations from the following year.

Hidden Coronal Mass Ejection (CME)

When we see the Sun in the sky (not directly, of course), we are seeing light originating from the Sun's surface, called the *photosphere*. This is the layer of the Sun that all the aforementioned astronomers were observing – a bright surface with dark sunspot regions. Although the photosphere is the Sun's surface, it is not the highest layer. Like the Earth, the Sun has an atmosphere; a tenuous, low-density region called the *corona*. (There is also another major region of the Sun between the photosphere and corona, called the *chromosphere*). The name corona comes from the Latin word for 'crown'. The word became famous in recent years due to the COVID-19 coronavirus, which also gets its name from the Latin word for 'crown', due to its crown-like appearance under a microscope. The corona of the Sun, the far superior corona, is not usually visible to us on Earth. The daytime sky is bright blue from scattered light from the Sun. Because of the bright blue sky, and brightness of the Sun's surface itself, we cannot ordinarily see the Sun's corona. This is also why we do not see stars in the daytime – they are dimmer than the sky. There is, however, one naturally arising situation in which we, using our naked human eyes, can safely witness the Sun's corona – during a total solar eclipse. Solar eclipses occur as the Moon passes between the Earth and Sun, completely blocking the Sun from view for a small region on the Earth's surface (we'll discuss this in detail later on). During the brief period where the Sun's surface is completely blocked (called *totality*), it is safe to look directly at the Sun (this is the only time it is safe to do this). During totality, the absence of the Sun's bright surface provides a rare view of the solar corona, wispy features emanating from the Sun into space. The image on the next page shows a sketch of the solar corona, as seen in the 1872 solar eclipse. Every total solar eclipse looks different due to the fact that the Sun's corona is continuously evolving over timescales ranging from hours to days. It is only in a total solar eclipse that the Sun's atmosphere is visible, as even a 1% sliver of the Sun's surface above the Moon is bright enough to hide the corona from us (and damage our eyes).

On 18th July 1860, a total solar eclipse was visible across a narrow path from Canada, across the Atlantic, through Spain and down to northern Africa. Much like other eclipses at the time, astronomers would sketch their view of the solar corona during the brief few minutes it was visible. The image on the next page shows such an image, sketched by G. Tempel from Torreblanca, in Spain. Most eclipse sketches, such as the image for the 1872 eclipse, convey the streamer-like features in the corona, which emanate radially away from the Sun. But in 1860, Tempel saw something different. In the bottom-right of his sketch, he drew a large spiral structure similar in size to the Sun itself. Although they would not be discovered until 1971, Tempel had unknowingly witnessed a coronal mass ejection (CME). A CME is an eruption of *plasma* from the Sun capable

Bright aurora, close to what miners in Colorado may have seen during the Carrington Event.

of travelling through the solar system towards Earth – a hidden link between solar activity and space weather effects (such as the aurora) on Earth. In the Science of the Sun chapter of this book, we'll discuss the science of solar eruptions in more detail.

Butterflies on the Sun

Although the discovery of coronal mass ejections was far away, the turn of the 20th century brought new scientific insights into sunspots. The two main players during this period were Edward Walter Maunder and Annie Russell Maunder. Before the pair were married, they both worked at the Royal Observatory, Greenwich, England (whose logo appears on the front of this book). Annie Maunder started at the Royal Observatory in 1895 and was one of the few female 'computers' at the time – a role that primarily involved doing calculations by hand for more senior staff. Walter (his preferred name) Maunder was also a staff member at the observatory, and the pair married in 1895. Unfortunately for them, due to civil service rules at the time, a couple could not both be employed by the civil service, and thus one of them needed to resign. Because Annie Maunder was more junior, with the lower salary of the two, she chose to resign. Thankfully, her work did not stop there. Annie continued to volunteer, and she and Walter continued to work on a number of projects, including eclipse expeditions, astrophotography, and sunspot science.

Annie and Walter decided to investigate the 11-year solar cycle, specifically how the location of sunspots evolved over this time. They meticulously combed through sunspot observations from 1877 to 1902 and created a plot of the date and latitude (north-south position) of every single sunspot to appear between those years. The plot they created, which would be published in 1904, is shown on the next page. Each line on the diagram is a sunspot, and if you use your imagination, you can see the shape of two butterflies in the data. For this reason, these types of diagrams (which are still constructed today) are called butterfly diagrams. Each 'butterfly' represents a solar cycle, lasting for around 11 years. At the start of the solar cycle, sunspots are primarily found at latitudes of 30° in both hemispheres. As the solar cycle continues, new sunspots begin to form at progressively lower and lower latitudes, increasing in frequency to solar maximum at average latitudes of 15°. By the end of the solar cycle, the number of sunspots is at a minimum,

Sketch of the 1872 eclipse.

Sketch of the 1860 eclipse, likely the first sighting of a coronal mass ejection.

and they form in a tight band around the solar equator at latitudes of 7° or less. As the next cycle begins, new sunspots once again form at higher latitudes, slowly beginning to ramp up in frequency for the next solar cycle. You can also see in the diagram that solar cycles are not self-contained. The older cycle can continue producing sunspots at lower latitudes, whilst other sunspots from the next solar cycle can occur simultaneously at higher latitudes. Previous astronomers, such as German astronomer Gustav Spörer, had already discovered a link between sunspot latitudes and the solar cycle, but it was the Maunders who popularised and pushed this theory further through their butterfly diagrams.

Another major discovery by the Maunders was the existence of a period of extended low solar activity from 1645 to 1715. The couple published their discovery in 1890 and 1894, building on previous work by Spörer. The Maunders verified that the absence of sunspots in solar data was in fact real, and not an artefact of limited observations from that period. This prolonged solar minimum, which was named decades later as the *Maunder minimum* in 1976, coincided with extreme low temperatures in Europe during the same period. It is a common myth that the Maunder minimum provides evidence that lower solar activity can create 'mini ice ages' here on Earth. Although a small number of scientists have advocated for this theory in the past, it is important to remember that correlation does not equal causation. Other theories suggest that the prolonged deep cold period was confined to Europe alone (not reflective of the global climate), or possibly linked to volcanic activity at the time. This is a complicated subject, as very little data from that time period exists.

As a woman, Annie Maunder played a role in these projects that was unjustly underplayed at first, and it would not be until 1916 that her

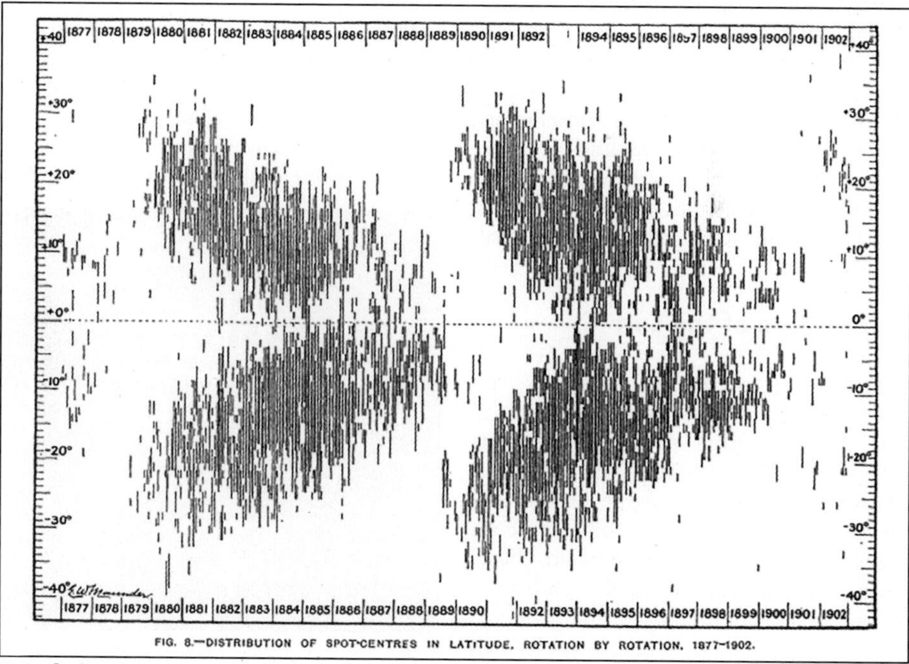

FIG. 8.—DISTRIBUTION OF SPOT-CENTRES IN LATITUDE, ROTATION BY ROTATION, 1877–1902.

Butterfly diagram constructed by the Maunders, published in 1904.

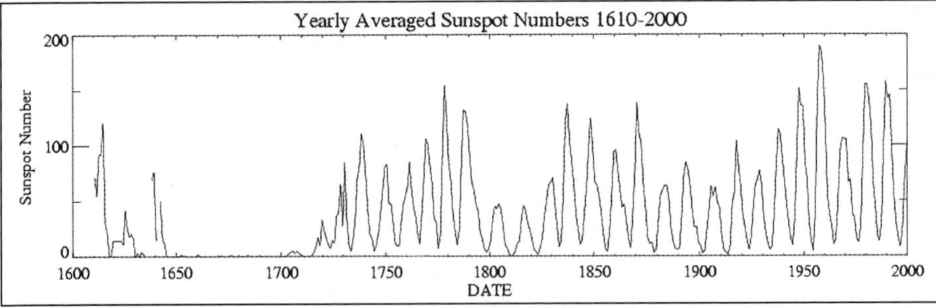

Sunspot numbers from 1610-2000, showing the maunder Minimum between 1645 and 1715.

contributions were finally fully appreciated. It was in this year that she was finally elected as a Fellow of the Royal Astronomical Society, less than a year after the block on electing women to the society was lifted. She would continue to remain a pillar of the solar physics community long after this, and for nearly 20 years after her husband's death, until her own death in 1947, at the age of 79.

Fast-forward to today, and Annie Maunder is recognised by the Royal Astronomical Society's annual 'Annie Maunder Outreach Medal', which is awarded to a recipient each year for outstanding outreach or public engagement in the fields of astronomy or geophysics. At the Royal Observatory, Greenwich, where Annie Maunder's career began, a new telescope was commissioned in 2018 to honour her life and achievements. It was named the *Annie Maunder Astrographic Telescope*.

True Nature of the Sun

Despite the significant progress on understanding sunspots between 1610 and 1904, the true nature of sunspots still remained a mystery. At this time, astronomers could not agree on whether these were caused by solar gaseous atmospheric effects (clouds), liquid effects (holes created due to meteor impacts), or surface effects (islands forming due to volcanism). This debate ended in 1908, when an American astronomer called George Ellery Hale finally solved the puzzle. At the Mount Wilson Observatory in California, Hale detected *Zeeman splitting* in light originating from a sunspot. It is not important for you to understand Zeeman splitting here (it is a complex and relatively uninteresting process in and of itself), but what is important is that Zeeman splitting occurs only in the presence of strong magnetic fields. That's right – sunspots are essentially giant magnets, with magnetic field strengths thousands of times stronger than the Earth's magnetic fields. George Hale had proved the existence of magnetism beyond the Earth, ruling it to be the fundamental force determining the evolution of the surface of not only our own Sun, but by extension all other stars too. This was a huge discovery, and the biggest of Hale's career. He, alongside a small number

English Heritage sign marking the old London residence of Walter and Annie Maunder in Lewisham.

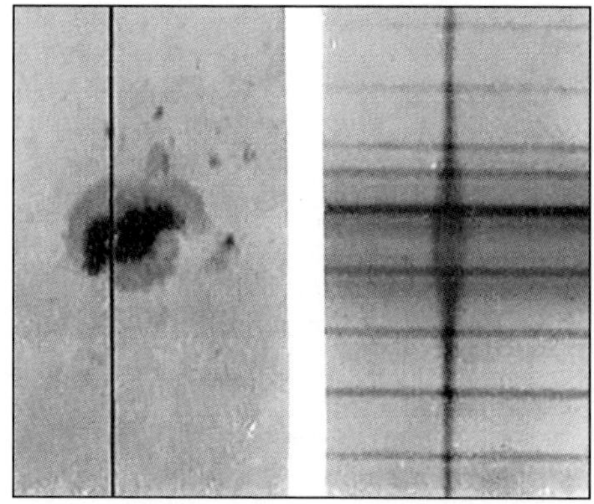
An early observation of Zeeman splitting in a sunspot.

of others, including Zeeman himself, had suspected that magnetic fields might be playing a role on the Sun, but Hale had proved it definitively. Due to their strong magnetic fields, sunspots repel the inflow of plasma from outside the sunspot. This causes the centre of a sunspot to become thermally isolated, and it cools to a lower temperature of 3,700°C (much cooler than the surrounding temperature of 5,500°C in the ambient Sun). The cooler temperatures of sunspots mean they don't glow as bright, which causes them to appear darker against the rest of the Sun.

Beyond sunspots themselves, Hale discovered the Sun has a global magnetic field too, which changes with the 11-year solar cycle. Building bigger telescopes to study magnetism on the Sun, Hale's discoveries continued. He found that individual sunspot regions can be approximated by dipoles, with both a north and south magnetic pole (like a regular bar magnet you may remember from school). In a specific hemisphere, the majority of sunspots are orientated in a certain direction, e.g. with the north pole on the right and south pole on the left. In the opposite hemisphere, sunspots are reversed, with the north pole on the left and the south pole on the right. This is known as Hale's law, and the image below shows a simple cartoon of this.

In another bizarre twist, with each new solar cycle, the direction of magnetic field in the sunspots of each hemisphere flips. This means that during one cycle, the northern hemisphere sunspots lead with a north polarity, only for the sunspots to switch to a south-leading polarity within the same hemisphere during the next cycle. This considered, the 11-year solar cycle is technically not 11 years at all, but underpinned by a 22-year cycle of varying sunspot numbers and magnetic field orientation.

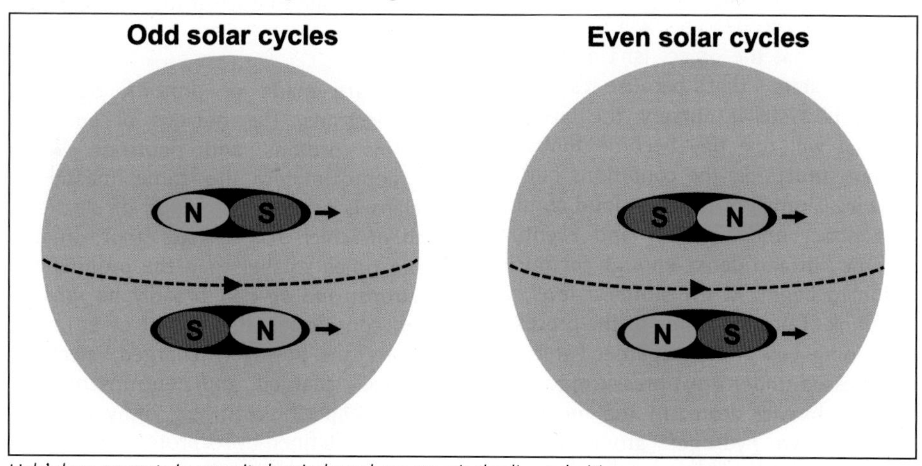

Hale's law: sunspots in opposite hemispheres have opposite leading polarities.

2: SCIENCE OF THE SUN

Life and Death of Stars

Let's take a step back for a moment, to about 5 billion years ago, long before humans observed the Sun. At this time, there was no Sun to observe. In the empty spaces between stars, there can sometimes exist huge clouds of gas and dust. After the invention of the telescope, these clouds were observed as faint, fuzzy objects, which astronomers named *nebulae*. Due to this, a nebula doesn't really mean a single thing in astronomy, as anything looking fuzzy through a telescope was given that name, long before astronomers knew exactly what they were looking at. Like all the stars in the sky, our Sun started its life as a small part of one of these ginormous nebulae of gas and dust. Due to gravity, parts of the cloud begin to contract. This happens very slowly, over tens of thousands of years, in multiple locations throughout the nebula. The most famous example of this kind of nebula today is the Orion nebula, located in the dagger beneath Orion's belt in the constellation of Orion. Orion's dagger appears to be made of three stars, but the middle star of the three is not a star at all – it is the Orion nebula. Looking at the object with a good pair of binoculars or small telescope, you will immediately recognise that the object isn't as sharp as the other stars – that's because it's not. The nebula is a stellar nursery, the birthplace of what will one day become thousands of stars, much like the conditions our Sun grew in. Clumps of the gas cloud continue to contract under gravity and eventually become hot and dense enough for nuclear fusion to begin. At the simplest level, you can think of nuclear fusion as the process of combining two atoms together (which are compressed under great pressure) to create a single heavier atom. In the most basic example, two hydrogen atoms combine to make a heavier helium atom. If the rate of nuclear fusion is high enough in the collapsing gas cloud, the process will release enough energy to create an outwards force, strong enough to resist further collapse of the gas cloud. As soon as nuclear fusion begins, the star is born. Throughout its life, the star exists as a constant battle between gravity trying to contract the star and nuclear fusion providing the force to oppose the star's collapse. This period of equilibrium between the two forces is called the main sequence period of the star's life and is the current phase of our Sun.

But how does this work? How can the combination of two atoms provide the energy needed to resist the collapse of a star? Well, the answer exists within what is perhaps the most famous equation of all time:

$$E=mc^2$$

This equation, first formalised by Albert Einstein, has a lot of applications, one of which happens within the core of the Sun. The equation states that energy (E) equals mass (m) multiplied by the speed of light (c) squared. What this means, in essence, is that mass and energy are interchangeable, but total mass-energy must be conserved between the two forms. How does this equation relate to nuclear fusion in the core of the Sun?

Atoms are made of protons, neutrons, and electrons. The nucleus of an atom contains protons and neutrons, which are approximately the same mass. The nucleus is orbited by a cloud of electrons, each of which has a mass approximately 2,000 times smaller than the protons and neutrons, and so can broadly be ignored when considering the weight of an atom. Protons are positively charged, electrons negatively charged, and neutrons have no charge. A specific atom, e.g. of hydrogen or helium, is defined by its proton number but can contain a different number of neutrons.

The Orion nebula, an example of where stars are born.

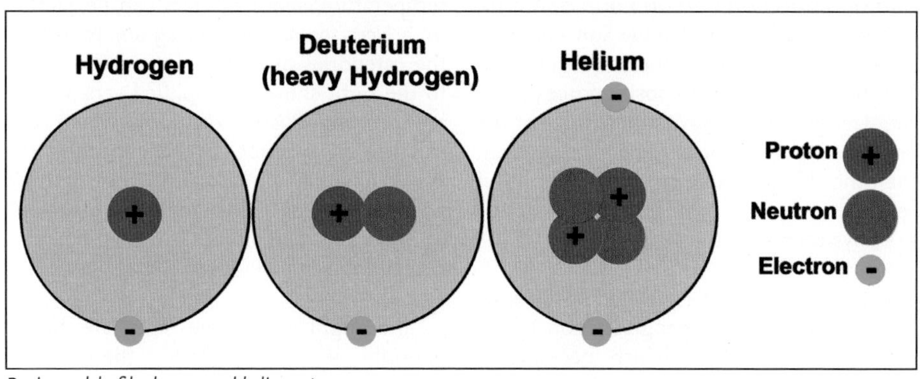

Basic model of hydrogen and helium atoms.

In the Sun's core there is both hydrogen and heavy hydrogen (also known as deuterium). They are both hydrogen, so contain one proton, but deuterium has one neutron to match, while regular hydrogen does not (it has no neutrons). The bulk of energy release from nuclear fusion arises from the heavy hydrogen, which has a mass of 0.000000000000000000000003344 grams. Helium, on the other hand, with two protons and two neutrons, has a mass of 0.000000000000000000000006645 grams. During the main nuclear fusion process, two heavy hydrogen atoms fuse together to create helium. (This is oversimplified, and there are additional fusion steps throughout the process, including the fusion of hydrogen to heavy hydrogen). If we sum their masses independently, the two heavy hydrogen atoms have a collective mass of 0.000000000000000000000006688 grams. For those with a keen eye, you'll notice that this mass is close to, but greater than, the mass of a helium atom. So where does this extra 0.000000000000000000000000042 grams of mass go when our hydrogen fuses into helium? The answer is in Einstein's famous equation. The mass doesn't disappear – it is converted into energy – the energy needed to power a star. This fusion process of hydrogen into helium is called the P-P chain (which stands for the proton-proton chain) and occurs around 10,000,000, 000,000,000,000,000,000,000,000,000, 000 times every second in the Sun – enough to resist the collapse of the star's gravity. If the strength of either opposing force (either the strength of the gravitational force or the energy output of nuclear fusion) decreases, then the star will expand or collapse until it reaches a new force balance. This will change the physical size of the star.

The colour of a star is related to its size. Massive stars have a larger gravitational force compressing the star, which necessitates a faster rate of nuclear fusion in the star's core to resist the larger gravitational force. This greater release of energy in the star's core makes it hotter. Although large stars have more hydrogen to fuse (they have more mass), the increased rate of fusion exceeds this fuel advantage, which means larger stars fuse through their available hydrogen more quickly. That is, the bigger they are, the brighter, hotter, and faster they burn. Smaller stars, on the other hand, despite having less fuel, have a far lower rate of nuclear fusion, which gives them far longer lives at lower temperatures. Unlike the colour convention of our taps, the hotter stars burn blue and the cooler stars red, with yellow somewhere in the middle. Blue being hotter than red may seem odd at first, but you can convince yourself that this is true from world experiences – a burning blue gas hob ring is far hotter than a yellow candle flame. The difference in life-spans between the largest and smallest stars is quite extreme. Large blue stars burn on the main sequence (the period during which they fuse hydrogen into helium) for a few 10-100 million years. Smaller red stars, on the other hand, can live for tens of billions of years. That's billions, with a B. Some small stars have lived for so long that their age nearly matches the age of the universe, around 14 billion years. Our Sun is a middle-aged, medium-sized star, about 4.5 billion years through its 10-billion-year lifetime as a main-sequence star. The difference in the brightness and temperature/colour of stars can be plotted on a *Hertzsprung–Russell diagram*, to explain the categories of different stars. The image to the right shows a simplified version of a Hertzsprung–Russell diagram, where you can see main-sequence stars forming the main diagonal component from bottom right (dim, cold, and red) to upper left (bright, hot, and blue). The diagram also displays the later stages of a star's life.

In the night sky, some stars are bright enough to distinguish the colour differences between them. For example,

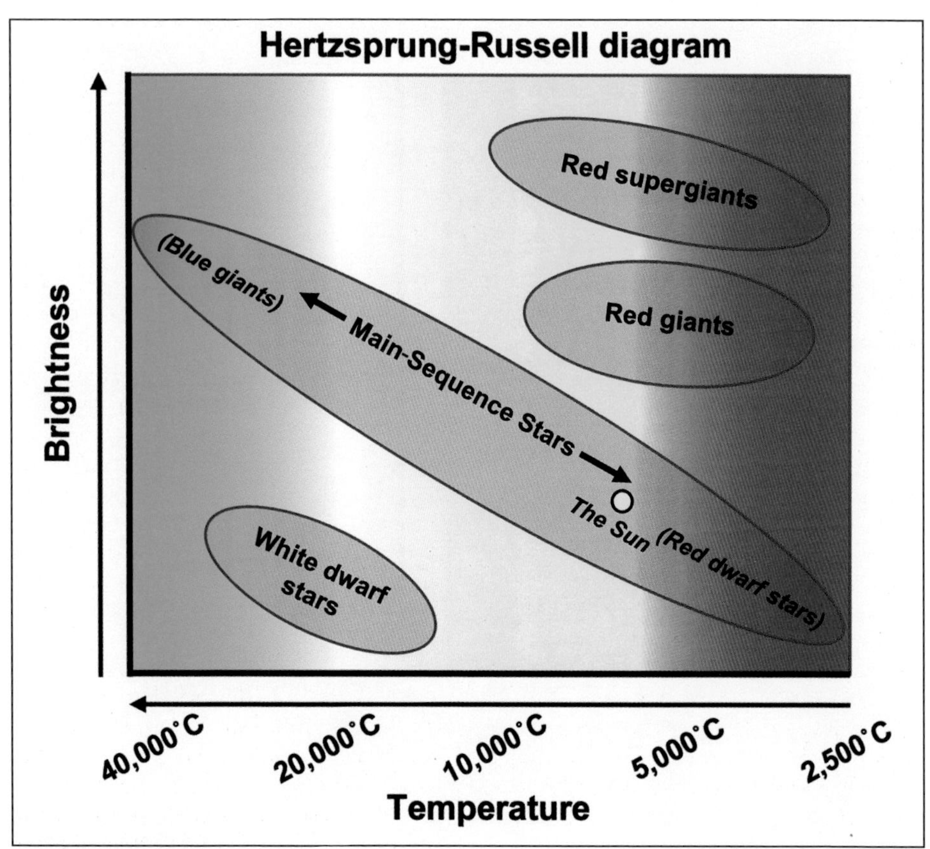

Simplified Hertzsprung–Russell diagram.

in the constellation of Orion, the unaided human eye can clearly see that the top-left star (Betelgeuse) is red and the bottom-right star (Rigel) is blue. By taking a long exposure photograph with a camera, you can better see the colour differences (between red, yellow, and blue) in many of the stars in the sky.

Unfortunately for stars, their time fusing hydrogen into helium as a main-sequence star does not last forever – eventually the hydrogen stockpiles in the core are exhausted. As the hydrogen begins to run out, the rate of energy release within the core slows down. Without the energy output from nuclear fusion, there is nothing to stop the collapse of the star from gravity. As the collapse begins and pressures and temperatures within the star rise, the core in the star's centre, where temperature and density are high enough to enable nuclear fusion, expands by a small increment. This is known as *ring fusion*, as fusion of hydrogen into helium can only occur in the outer edge of the newly grown core, where the hydrogen has not yet been depleted. Fusion in the core ring can occur in incremental steps, as the process repeats several times to expand the core size marginally. But ring fusion does not last forever, and eventually there is no more available hydrogen within the core's maximum size. Hydrogen still exists in the outer layers of the star, but fusion can only occur in the star's core. With the absence of fusion once more, the star will continue

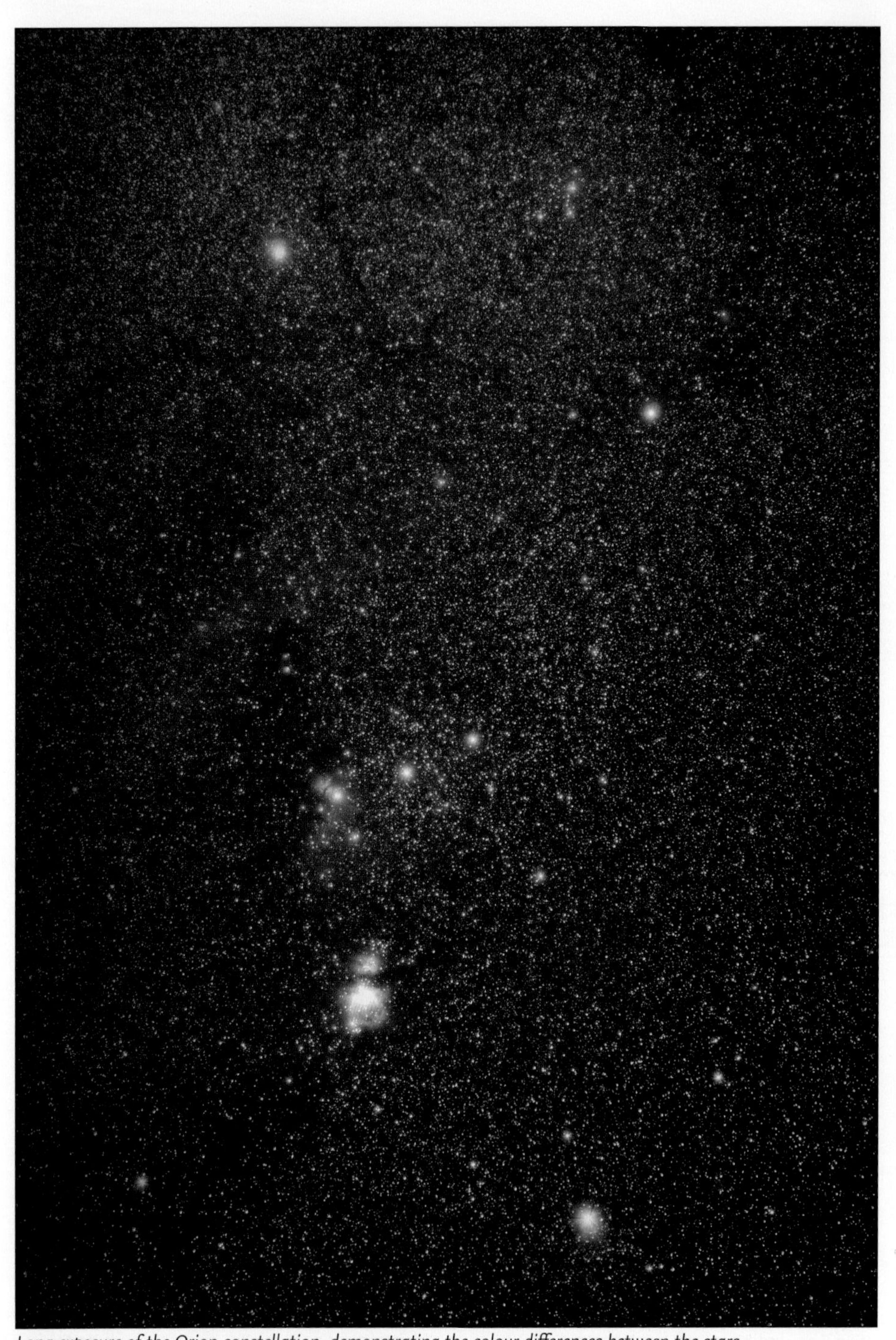

Long exposure of the Orion constellation, demonstrating the colour differences between the stars.

to collapse under its gravity, but things are not over yet. Without hydrogen burning to resist the collapse, the gravity continues to compress the star further. This collapse continues to heat up the star to pressures and temperatures far exceeding what was initially required to kickstart hydrogen burning. If the star has enough mass to compress itself sufficiently, conditions may be right for the nuclear fusion of helium into carbon. This interaction releases far more energy than the fusion of hydrogen to helium and causes the power output of the star to increase dramatically. This new level of nuclear fusion provides enough force to resist the gravitational collapse, and much more. The increased power output in the core pushes against gravity, expanding the star to many times its initial volume. The star gets so large that the outermost layer cools down due to its distance from the power source in the core. This causes the star to burn a cooler red colour – becoming a red giant. This will one day happen to the Sun. The Sun will expand so greatly that it will engulf the orbits of Mercury and Venus, reaching a final size almost as large as Earth's orbit. The Earth is 100 Sun widths from the Sun currently, which means the Sun will expand by nearly 100 times its current size. Even if Earth escapes the Sun's expansion, its proximity to the new Sun surface will certainly render it uninhabitable. This won't happen to the Sun for a few billion years though, so I wouldn't lose any sleep over it.

As you may have already deduced, the nuclear fusion of helium into carbon cannot continue forever either. Stars in this phase burn much faster, so spend significantly less time as a red giant than they did in the main sequence. The next life step of a star depends entirely on the nature of the star itself. If the mass of the star allows, the end of helium fusion into carbon could be followed by the fusion of elements with progressively higher proton numbers (this concept is explained later), such as carbon into oxygen, oxygen into nitrogen, and so on. This is around the limit of our own Sun, which, unlike larger stars, does not have enough mass to provide the gravitational force necessary to enable nuclear fusion beyond these elements. After this point, nuclear fusion will end for the last time on the Sun, and our star will no longer be capable of resisting the inevitable collapse of itself under its own mass – a battle it will fight for the next 10 billion years. As the collapse continues, the atoms within the star are pushed together so tightly that the entire Sun is crammed into a volume about the size of Earth. For Sun-sized stars and smaller, this collapse halts when the atoms themselves cannot be forced together by gravity alone; the collapse is resisted by a force called *electron degeneracy*. The leftover

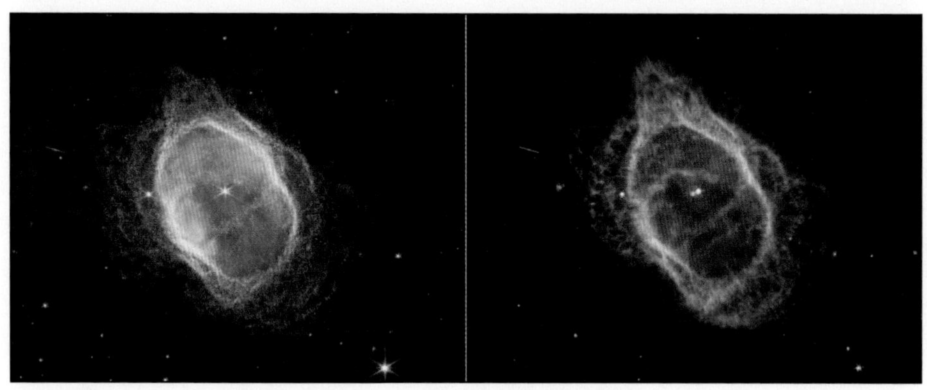

Ring Nebula observed by the NASA JWST. Our own Sun shares a similar fate.

star has no energy source, but remains a hot, dense object called a *white dwarf star* and takes billions of years to cool down. During the process of the star collapsing, outer layers of the star may untether from the shrinking core, and are left to drift out into space. This leftover gas is illuminated by the white dwarf, forming a beautiful cloud of striking colours called a *planetary nebula*. You may recall that nebulae were classified due to their fuzzy appearance in a telescope and not by their actual nature. This is why the birthplace and death place of stars are both called nebulae. The Sun will one day collapse to a white dwarf, with a planetary nebula expanding through where the planets exist today. The image on the previous page shows a famous example of a planetary nebula, called the Ring Nebula, as seen by the NASA James Webb Space Telescope (JWST). The white dwarf star is visible in the centre of the nebula.

Stars much larger than the Sun will not become a white dwarf star and ring nebula. The largest red giant stars, called *red supergiants*, have the gravitational force needed to fuse elements heavier than nitrogen, fusing through every element up to iron. This is how all elements in the universe lighter than iron (and heavier than helium) were formed, via nuclear fusion within the core of a red giant star. When large stars attempt to fuse iron, something different happens. Up to this stage, the fusion of two elements together releases energy. But not iron. As the nuclear fusion of iron takes place, the process uses more energy than it releases. This nuclear reaction, therefore, creates an energy deficit, no longer producing the energy needed to oppose the gravity of the star. For all stars in the iron-burning stage, an energy void forms in the star's core, and the collapse is accelerated. The force of the imploding star smashes through the resistance of electron degeneracy, the stage that resists the collapse of Sun-sized and smaller stars. The critical mass at which this threshold is passed is called the Chandrasaker limit, named after the Indian astronomer who developed the theory. This limit occurs in any remaining star core with a mass of over 1.44 times the mass of our Sun. The implosion produces an unfathomable amount of energy, and the collapse of a dying star produces one of the most energetic events in the galaxy – a type II supernova. During the supernova, all elements heavier than iron are produced. Gold, silver, platinum, etc. – they were all formed within a dying star. After these elements are expelled from the star during the supernova, they eventually form a cloud of gas and dust. This gas cloud will eventually collapse again under gravity to create a new generation of stars and planets. This cycle of stellar life and death repeats itself, and our Sun is most likely a third-generation star formed partly from material left behind by previous stars.

What remains of the star after the supernova depends again on the star's mass. After passing the limit of electron degeneracy, the atoms are compressed together until a bizarre quantum physics effect happens. Negatively charged electrons and positively charged protons are forced together so tightly that they combine into a neutrally charged neutron. The compressed neutrons can resist the star's collapse to a certain point, forming a super dense object called a neutron star. Atoms are mainly empty space, so by compressing a star to its neutron scales (which is known as *neutron degeneracy*), the remaining object is many times denser than white dwarf stars, which are propped up by resistance on atom scales (electron degeneracy). Neutron stars can contain the entire mass of the Sun in a sphere about 20 km across.

However, even neutron degeneracy has a limit to what it can resist. That limit occurs when the remaining star core (after many layers have been blown into space by the supernova) contains a mass of over 1.44 times the mass of our own Sun. Beyond this mass, gravity is too strong for even neutron degeneracy to resist. After this, there are no more processes to resist the gravitational collapse any further. The mass of the star will then collapse infinitely, until the entire star's mass exists within a one-dimensional, infinitely small point known as a singularity. Our current understanding of physics begins to break down within these singularities, but we know they have a finite mass (determined by the mass of the star forming it, plus anything that falls in afterwards) within an infinitely small area and infinitely large density. The gravitational field of the singularity is so dense that not even light can travel fast enough to escape from a finite distance away, determined by the singularity's mass. This region where light cannot escape is a *black hole*, and the edge of the black hole (the limit of where light can escape) is called the *event horizon*.

On its current life trajectory, our own Sun will never become a black hole or neutron star – it is simply not big enough to surpass the Chandrasaker limit. Although the Sun's evolution into a red giant and white dwarf star is billions of years in the future, the Sun also varies on much smaller timescales (albeit much less dramatically). Although nuclear fusion is the driving force within the Sun, we see little influence of it on the Sun's

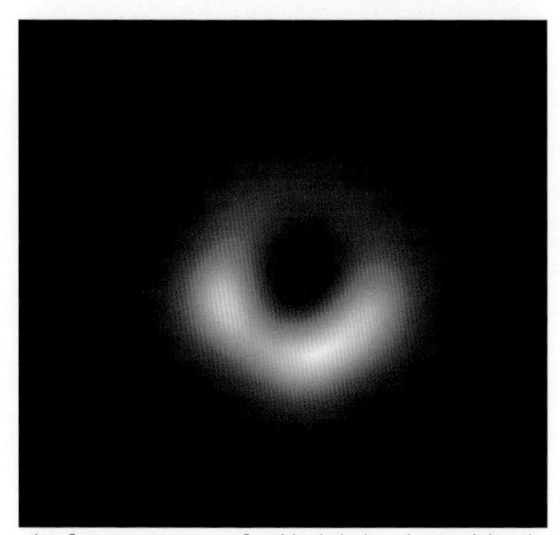

The first ever image of a black hole, observed by the Event Horizon Telescope in 2019.

surface. Activity on the Sun's surface and within its atmosphere, such as sunspots, the solar cycle, and solar flares, is at the whim of a different fundamental force – magnetism.

Forces of Nature

In nature, there are only four primary forces determining the universe around us. These forces are completely independent of one another and are known as the fundamental forces. These forces are *gravity*, the *weak nuclear force*, the *strong nuclear force*, and *electromagnetism*. Any other forces you've heard of are not independent forces and fall under the umbrella of one of the four fundamental forces. Gravity is the force of attraction between any objects with mass. The strength of the gravitational force between two objects depends on only two factors – the mass of each object and the distance between them. Gravity works over a huge range of distances; it is the same gravity keeping you tethered to the ground and dictating the orbits of the planets around the Sun and the Sun around the centre of the Milky Way Galaxy. Not only is the Earth's gravity pulling you towards it, but your gravity is also pulling the Earth towards you. The problem is that your gravity is irrelevant compared to the Earth's, so has no effect. This is not true with larger objects, however. Two objects of a similarly large mass would not technically orbit each other, but instead both orbit the 'centre of gravity' between them, called the *barycentre*. For a complex system of multiple gravitational bodies, such as the solar system, all the objects orbit around the common centre of gravity, which is determined by the mass and distance of everything in the solar system from one another. You may think that the barycentre of the solar system is at the centre of the Sun, but this is not always true. With a mass of only 1/1000 of the Sun's mass, Jupiter is large and far enough away from the Sun to skew the solar system's centre of gravity towards it. Every other planet, dwarf planet, comet, and asteroid has some effect

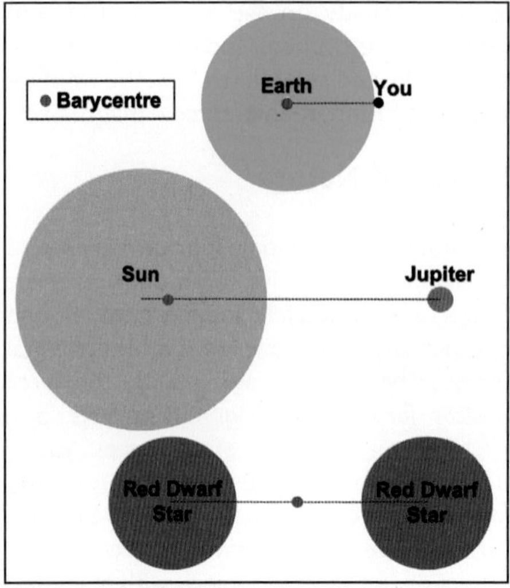

The centre of gravity between two objects.

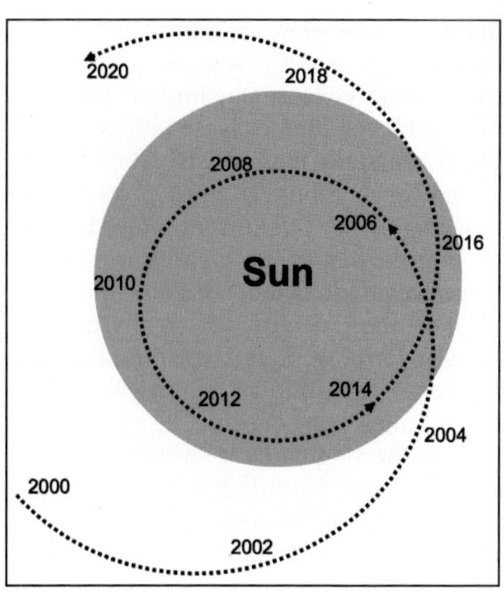

Evolution of the barycentre of the solar system.

on the barycentre too, but not as much as Jupiter. The image above shows the evolution of the solar system barycentre, which is sometimes outside of the surface of the Sun. The Sun orbits this location too, which causes the Sun to appear to wobble. Detecting this wobble in distant stars is another method of discovering exoplanets, as it implies that the star's barycentre is skewed by the presence of a large planet.

The second fundamental force is called the weak nuclear force. We have already discussed this force, as it is the force behind nuclear fusion. It is also responsible for *nuclear fission*, the opposite of nuclear fusion. Nuclear fission occurs as very heavy unstable elements decay to a lighter one, releasing energy in the process. Elements lighter than iron will release energy through nuclear fusion, whilst elements heavier than iron require energy for nuclear fusion (and so can only happen in a supernova). If we flip this around to the context of nuclear fission, elements heavier than iron will release energy when decaying via fission to a lighter element, whilst elements lighter than iron would require a lot of energy to force fission to happen. In nature, most elements are fairly stable and do not spontaneously decay via fission to a lighter element. The common exceptions to this are found at the end of the periodic table, in heavy and unstable elements like uranium. Nuclear fission of these elements is used for the production of energy in nuclear power plants. Power production via fission is very efficient, producing a lot more energy per tonne of uranium compared with the burning of carbon-based fuels such as coal or oil. The caveat is that you're left with radioactive by-products which cost a lot to dispose of in a safe way. We do not yet have the technology to produce energy via fusion in the way we do fission, using the same mechanism fuelling the Sun, which would create clean energy without radioactive by-products. There's a chance this new, clean energy solution will arrive in the next decade or so, but we're not there yet.

The next fundamental force is the strong nuclear force, and it is the hardest of the four to conceptualise. Although this force is similar in name to the weak nuclear force, it is completely different (as are all the fundamental forces). Whereas the weak nuclear force holds protons and neutrons together in an atom (releasing/absorbing energy as the configuration of the atom changes), the strong nuclear force is the force holding together an individual proton or neutron. We won't be exploring this in this book, as sub-particle physics has little to do with the day-to-day physics of the Sun. To provide a very quick summary – protons and neutrons are made of fundamental particles called *quarks*. Quarks are the building blocks for protons, neutrons, and other exotic particles, held together by the strong nuclear force. Both the weak and strong nuclear forces play a role in the death of giant stars. Electron degeneracy, which stops a white dwarf from collapsing, is resisted by the weak nuclear force (the resistance between atoms). Neutron degeneracy, however, resists the collapse of neutron stars via the strong nuclear force (the resistance between neutrons).

The final fundamental force is electromagnetism. Electromagnetism consists of two components – electric fields and magnetic fields. From our everyday experience, electric and magnetic fields may seem like separate things. Electric fields allow the flow of electricity through our cables, power lines and electronic devices, whereas magnetic fields exist in ... magnets? In reality, these forces are two aspects of the same thing, electromagnetism. Variation in an electric field will produce a corresponding magnetic field, and, similarly, a change in a magnetic field will produce an electric field. In both cases, the new induced force is perpendicular to the direction of the time-varying force. This is the principle of electromagnetism. Planets and stars produce their magnetic fields via electromagnetism. Inside the Earth, a layer of liquid iron rotates around a solid inner core, carrying with it an electric field. This rotation of the electric field produces a magnetic field large enough to extend far above the Earth's surface. The same mechanism produces Jupiter's magnetic field, except with a layer of metallic hydrogen instead of iron. In the Sun and other stars, the rotation of hydrogen plasma produces the magnetic field.

The Fourth State of Matter

As a child in school (at least in the UK National Curriculum), you are taught about the Three States of Matter – solid, liquid, and gas. However, these categories exclude some 99.9% of the matter in the observable universe. The Sun, stars, and near-empty spaces between them are made up of the fourth state of matter, plasma. The transition from solid to liquid, and from liquid to gas, comes by adding energy into the material. In a crude summarisation, extra energy added to a material causes the atoms within it to vibrate faster (just as you move faster with more energy). Eventually, after the addition of enough energy, the atoms within the material can no longer be bound to their original state. At this stage, atoms in a solid configuration would transition to a liquid configuration (the solid would melt), or atoms in a liquid state would transition to a gaseous state (the liquid would evaporate). In some materials or under certain conditions, it is possible for a solid to skip the liquid stage and transition directly from a solid to a gas. This process is called sublimation. Removing energy from the atoms will reverse the process, causing condensation and freezing. Energy can be added to atoms in two main ways, by increasing either the temperature or the pressure of the atoms' environment. For example, most people know that you can boil liquid water into gas by raising the temperature to 100°C. But the boiling point of water is only 100°C at sea level. At altitudes of 14,000 feet, the air pressure is lower and water boils instead at 85.5°C. Therefore, for 86°C liquid water at sea level, you would only need to do one of the following to boil it – increase the temperature or decrease the pressure.

Plasma is created by adding further energy to the atoms. Like the transitions from solid to

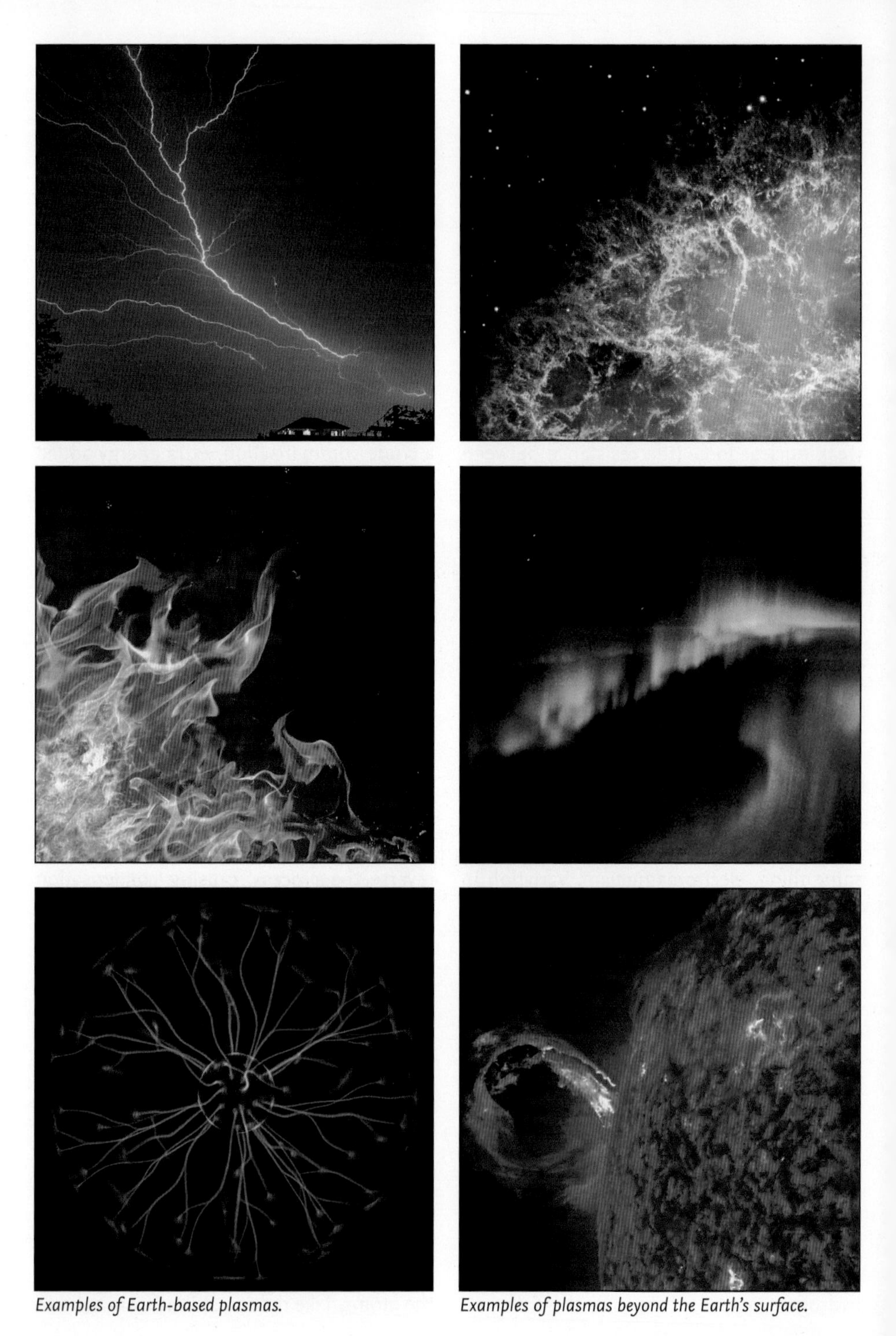

Examples of Earth-based plasmas.

Examples of plasmas beyond the Earth's surface.

liquid to gas, the next stage is the transition to plasma. Plasma does occur naturally on Earth, but requires high energies. Fire and lightning are both made of plasma, for example. Despite its name, blood plasma is not plasma. Beyond the Earth, plasma is abundant throughout the universe. Plasma is the main ingredient of the Sun, made up of, primarily, hydrogen (80%) and helium (20%), but with trace amounts of many heavier elements including carbon, calcium, silicon, oxygen, and iron. Equivalently, stars are also made of plasma, as are the spaces in between them. The atmospheres of some planets, including the upper atmosphere of the Earth, are made of plasma too.

Most of the immediate world around us is neither positively or negatively charged – it is neutral. A neutral substance, whether it be a solid, liquid, or gas, has an equal amount of protons (which are positively charged) and electrons (negatively charged). In an atom, the nucleus contains protons and neutral neutrons surrounded by a cloud of electrons. By adding enough energy to a neutral gas, electrons bound to the atom can escape. This transition results in an atom that is now positively charged (it has more protons than electrons) and a free electron (still negatively charged). An atom missing one or more electrons is known as an *ion*. This is plasma, a fluid of ions and free electrons. Despite its charged components, plasma is neutral on large scales. This is because, although the ions and electrons are charged, there is an equal abundance of them across the whole plasma. Due to the charged nature of its components, plasma has a unique relationship with electromagnetic fields. The study of how plasma behaves in a magnetic field has an excellent name – *magnetohydrodynamics*. Magnetohydrodynamics determine the behaviour of the solar atmosphere.

Ions have different levels, depending on how many electrons have been lost. Hydrogen is the lightest element, with one electron and one proton, and becomes fully ionised after the loss of this single electron. Heavier elements, however, have many more electrons to lose. Take an iron atom, for example (denoted Fe in the periodic table), which has 26 electrons in its neutral state. Neutral iron is the default version of iron, which we can call Fe I. After the atom gains energy and loses its first electron, it becomes singly ionised (it's lost one electron), which we call Fe II. Fe II becomes Fe III after another electron loss, continuing all the way to Fe XXVII with no electrons left. If energy were removed from the plasma, ions would recapture the electrons one by one in a process called recombination. Increasing amounts of energy are required to strip each additional electron – the hotter the plasma, the higher the ion. Iron ion levels up to Fe XIV are abundant in the hot solar atmosphere, but fully ionised Fe XXVII can only exist under rare, extreme conditions on the Sun – during the largest solar flares.

Colour of the Sun – More than Meets the Eye

What colour is the Sun? The question seems simple enough, but it is one without a simple answer. Before we begin, let's quickly discuss the basics of light and colour. Light is a wave of *photons* (little packets of energy) travelling in a vacuum with a constant speed of 300,000 km/s – the speed of light. Through thicker mediums, such as air or water, light will travel slightly slower than this. Much like waves in the ocean, light waves have a wavelength, the peak-to-peak distance of a wave. Light is far more expansive than the light visible to our eyes. The light visible to humans spans a narrow range of wavelengths covering 400–700 nm – a range of light we call visible or optical light. One nm, or nanometre, is a million times smaller than a millimetre. The wavelength of a light wave determines its colour. The 400–700 nm range of

A naturally occurring spectrum – a rainbow.

visible light spans from violet light, at the shorter wavelengths, to red light, at longer wavelengths. White light is a combination of this entire visible range and can be split up into the red to violet colour continuum. The splitting of light into different wavelength (or colour) components is called a spectrum, as happens naturally during a rainbow.

Beyond what the human eye can see, there are other types of light. At wavelengths shorter than violet light, we reach ultraviolet light, or UV. The energy of a light wave depends on wavelength and is larger for shorter wavelengths of light. UV light, therefore, has a higher energy than visible light, which makes ultraviolet harmful to our skin (whereas optical light is not). At shorter wavelengths (and higher energies) than ultraviolet, we reach X-rays, followed by gamma rays at the very shortest wavelengths. With their higher energies, X-rays are even more harmful than UV light, and exposure to these should be kept at a minimum (as is necessitated by the health benefits of using a medical X-ray) over a lifetime. Gamma rays however, should be avoided at all times, with the exception of certain cancer treatments. At wavelengths longer than visible light, we have more different types of light. Increasing wavelength beyond red light gives us infrared, followed by microwaves, and finally, radio waves at the longest wavelengths. Collectively, this range of light is known as the electromagnetic spectrum of light and consists of (in order of increasing wavelength and decreasing energy) gamma rays, X-rays, ultraviolet, visible light, infrared, microwaves, and radio waves. Different types of light have versatile uses for humans, but in terms of the physics, they are all the same thing – light. The only difference between a radio wave and a gamma ray is the wavelength of the light wave (and, therefore, the energy the wave carries).

The human eye sees colour based on what wavelengths of light an object emits or reflects. For example, grass absorbs most of the visible light spectrum, reflecting only green light, and so appears green to our eye. An orange neon sign, on the other hand, is emitting exclusively at an orange wavelength, so we see it as orange. However, most emitters of light emit at more than one specific wavelength; they emit a range of light with a peak wavelength dependent on the temperature of the object. If an object is close to being a perfect emitter, like a star, then its emission follows a curve called a *blackbody* function. All this function is is a relationship between an object's temperature and its brightness at different wavelengths. The curve has

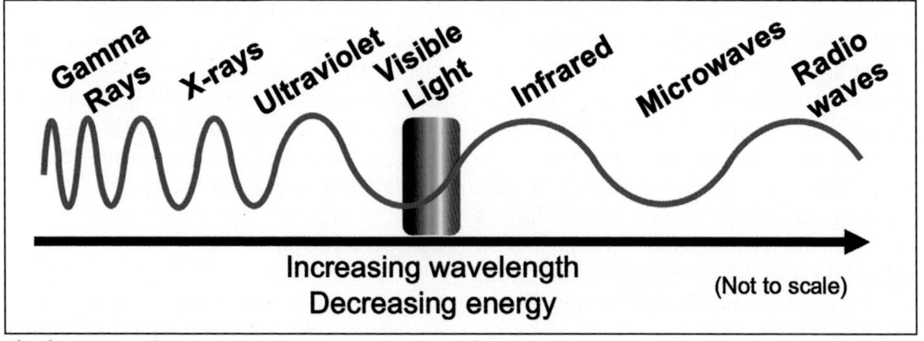

The electromagnetic spectrum.

a peak, but with emission across a wider range of the spectrum. Hotter objects will have a shorter peak wavelength, and cooler objects a longer one (as seen in the graph below). This manifests itself as something we've already discussed, the fact that hotter objects burn blue and cooler objects burn red (e.g. blue stars are hotter than red ones, a blue flame on a gas burner is hotter than a yellow candle). Humans are heat sources too, so also emit at a peak wavelength. Obviously, humans don't glow in any light visible to us, and this is because our peak emission is in the infrared, a 'colour' of light we cannot see. Thermal imaging cameras work by measuring this infrared light, in

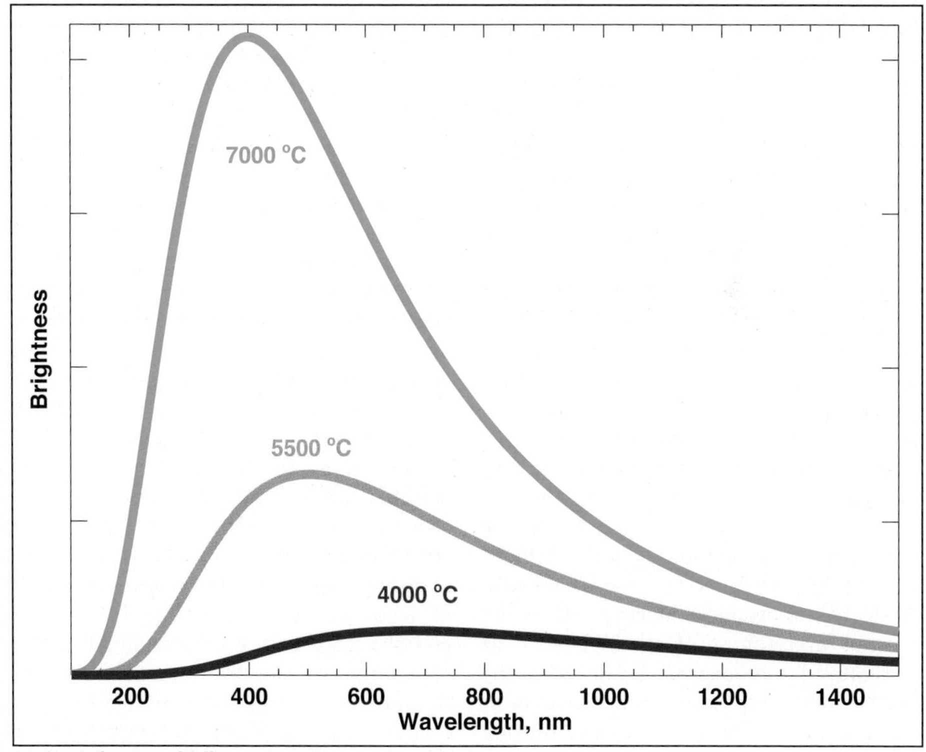

Blackbody function of different temperatures.

which humans glow up like Christmas trees. Infrared has a longer wavelength than visible light and, therefore, a lower energy. Shorter wavelengths have higher energies, which explains why blue light emits from a higher energy source than red light. This blackbody rule only applies for perfect emitters, which are objects whose light is produced by its heat alone. (Purpose-built lights you use at home create their colour using a mechanism other than heat).

The Sun, like other stars, can be approximated as a blackbody emitter. With a surface temperature of 5,500°C, the Sun's emission peaks at a wavelength of 480 nm. In the visible spectrum, 480 nm corresponds to a colour of blue-green. Huh? The Sun is blue-green? Well, not really. 480 nm is the peak of the Sun's emission, but the Sun emits brightly over all visible wavelengths of light. Our eyes see all of these wavelengths mixed together, creating white light, the true colour of the Sun. If the Sun is white, where does the idea of a yellow Sun come from? As sunlight passes through the Earth's atmosphere, some of it is scattered. Shorter wavelengths, at the blue end of the spectrum, are easier to scatter, giving us a blue sky. The sunlight that reaches us is then missing some of the blue light it started with, skewing the white light towards yellow. When sunlight has more atmosphere to pass through, for example, when it's low on the horizon at sunrise or sunset, longer wavelengths of sunlight can be scattered along with the blue, giving the sky a yellow/orange appearance, leaving only the red sunlight to reach our eyes – giving the Sun a red colour. Sunlight is more than just the colours of the rainbow, though, and includes types of light we cannot see too. Thankfully, the Earth's atmosphere blocks most of the harmful wavelengths of sunlight, such as UV light. If the atmosphere did not block the worst of UV light, sunlight would've been too harmful for life ever to evolve outside the ocean.

The main spectrum of sunlight, which is determined by its temperature via the blackbody function, is called the *Sun's continuum*. As I'm sure you'll be pleased to know, there is more to the Sun's spectrum than just the continuum – the story of sunlight does not end here.

Spectroscopy

Spectroscopy is the study of spectra (the plural of spectrum). Before we continue, let's recap two concepts. Firstly, it is important to remember that the energy of a light wave is determined by its wavelength, with higher energies at shorter wavelengths. Secondly, we've learnt that plasma is a fluid of ions and free electrons, and that the level of an ion (how many electrons it has lost) depends on the energy of the plasma. As an electron escapes from an atom or changes energy level within that atom, it requires a very specific energy to do so. Let's combine these ideas. If you were to take a tank of hydrogen plasma and shine white light through it, most of the white light (ranging from red to violet wavelengths) would pass straight through. However, one very specific wavelength of light, the one carrying the specific energy needed to excite the electron within the hydrogen atom, would be absorbed by the electron. The electron would absorb this specific wavelength of light, taking the energy for itself. If this absorption by electrons occurred enough times within the plasma, then the spectrum of white light would be missing the colour corresponding to this specific wavelength (which itself matches the specific energy absorbed by the electron). The absence of certain wavelengths of light is called an absorption spectrum. The main wavelength of light absorbed by hydrogen is called H-alpha at 656.3 nm, corresponding to a slice of red light missing from the solar spectra. Because the energy levels of electrons in each atom are different, so are each of the wavelengths of light absorbed by each atom.

This means that, an absorption spectrum acts as a kind of signature or barcode unique to its source. By studying the Sun's spectrum, the absence of specific wavelengths of light indicates which atoms are present on the Sun, because those atoms have absorbed the missing light. The Sun has many absorption lines, which allows us to determine its composition – mainly hydrogen and helium, with traces of many other elements too. In fact, this is how helium was discovered. In 1868, gaps in the Sun's spectra were observed that couldn't be explained by any other element. Astronomers determined that it must be something new, and so they named the element after the Greek word for Sun – Helios. Helium wasn't discovered on Earth until some years later in 1882, in gases emitted from the top of Mount Vesuvius. The image below shows the full visible spectrum of the Sun, each dark line marking absorption from plasma in the Sun. Some of these absorption lines are strong (can you see the large absorption from H-alpha in the red?), and others are weak. The absorption lines also extend beyond the visible range into infrared and UV light. You have to be careful when examining the solar spectrum, however, as some absorption lines are not from the Sun at all, but from oxygen and nitrogen in the Earth's atmosphere. We call these specific absorption lines telluric lines, Latin for 'of the Earth'. Spectral observations from space need not worry about telluric lines.

In plasma, electrons are not only energised by incident light. If densities and temperatures are high enough, collisions in the plasma can provide the energy needed, converting the kinetic energy of the plasma to the energy of the electrons. The energisation of electrons in this way does not absorb any light, so does not produce absorption in the solar spectra. This process of collisional excitation is common in the atmosphere of the Sun (the corona), where temperatures are much higher than on the surface.

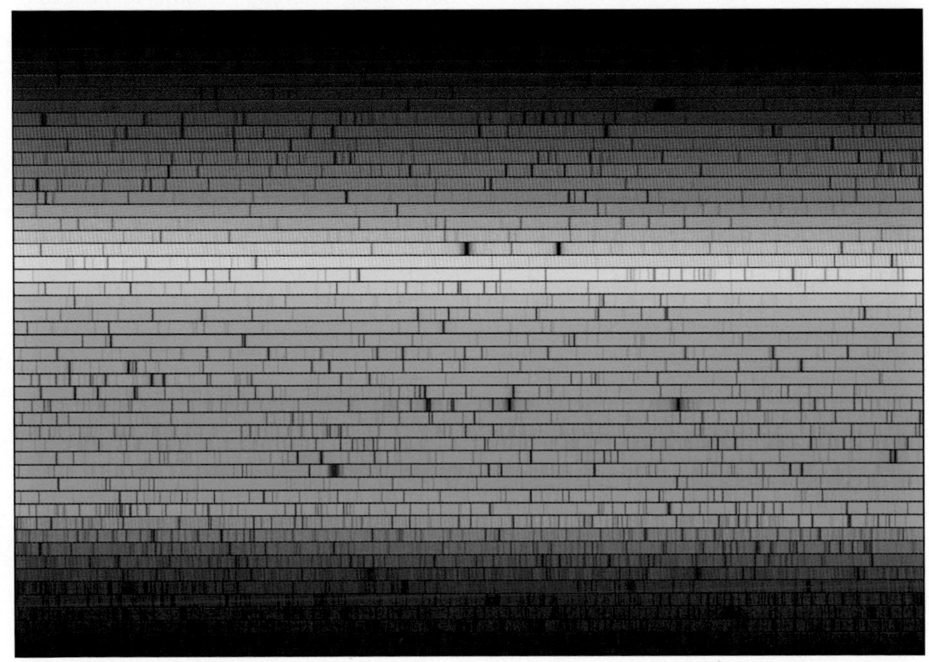

The visible spectrum of the Sun.

Once an electron gains the extra energy via a collision, it does not keep it forever. Eventually the electron will want to dispose of the extra energy, to return to a more relaxed state. As it does so, the energy must go somewhere. The energy is released as a photon of light emitted at the wavelength equal to the energy released by the electron. As the emission at a specific wavelength occurs many times within the plasma, the direct inverse of an absorption spectrum is created – an emission spectrum. The image below shows a comparison between an absorption spectrum and emission spectrum. The absorption spectrum shows the full colour range, with gaps marking absorption, whereas the emission spectrum shows colour only where light is emitted. As you can see, the wavelength is the same for both types of spectra, depending on the unique energy levels of each ion. Absorption is more common on the Sun's surface, whereas emission is most common in the Sun's atmosphere.

By measuring the strength, thickness, and position of the lines in the spectrum, we can learn a lot about the plasma creating the absorption or emission. A specific ion level can only exist in a finite temperature and density range. Take Fe XII for example, an iron ion with eleven lost electrons. If temperatures are too low, the ion does not have the energy to lose 11 electrons. If temperatures are too high, the ion will lose far more than 11 electrons, and Fe XII cannot exist. With the knowledge of what conditions an ion can exist in, by studying the strength of the spectral line of a specific ion relative to others, we can calculate the temperature and density of the source plasma.

The position of the spectral line can also reveal the velocity of the source plasma. If you cast your mind back to school, you may recall learning about *Doppler shift*. Doppler shift occurs when the wavelength of a wave appears different for both a moving source and stationary observer. If we use sound waves as an example, a stationary car horn will sound at a certain pitch (the wavelength of the sound wave). If the car is moving when the horn sounds, passengers within the car will hear the horn at its original pitch. Observers outside the car will hear something different. For observers ahead of the moving car as the horn sounds, sound waves produced by the horn compress as the car moves in the same direction. For observers behind the car, the sound source is moving away from them, so waves emanating from the car are stretched from

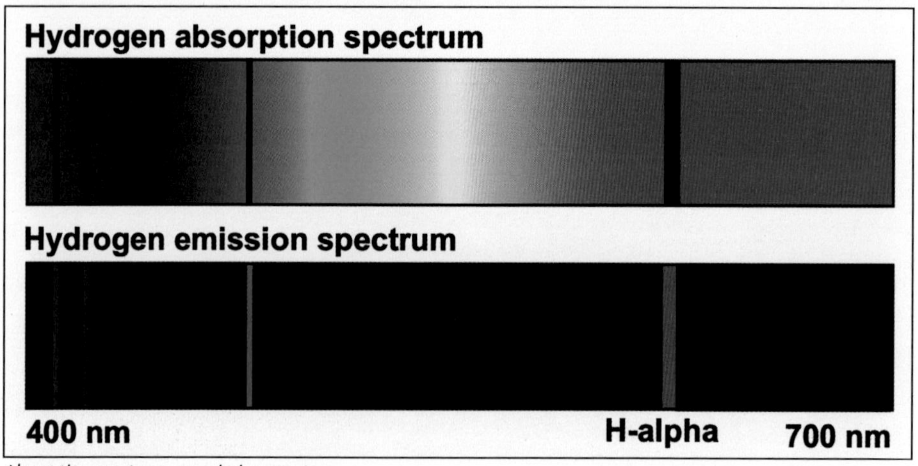

Absorption spectrum vs emission spectrum.

the observers' perspective. The compression and extension of the sound waves produce a change in the pitch heard by the observers. The image below provides a diagram of this. The Doppler shift applies to any moving wave source. Imagine a fast car or sirens speeding by, the iconic change in pitch of sound as it passes by is caused by the rapid compression and extension of sound waves. This effect is called Doppler shift, first discovered in 1842 by Austrian physicist Christian Doppler.

Doppler shift also applies to light waves. Spectral lines are produced at a specific wavelength. If the source plasma of the spectral line is receding from the observer, the wavelength is stretched, which shifts the spectral line to a longer wavelength. This increase in wavelength is called red shift, as the line position moves towards the red end of the spectrum. Alternatively, if the source of the spectral line is moving towards the observer, light waves are compressed and observed to shift towards the blue end of the spectrum – blue shift. And thus, by comparing the expected and observed wavelength of a spectral line, you can calculate the velocity of the light source. Doppler shift only reveals information on the velocity in our line of sight, and not of motion within the plane of the sky. The astronomer Edwin Hubble used Doppler shift to measure the velocity of galaxies, providing key evidence for the big bang by discovering that all galaxies move away from one another. On the Sun, we can similarly use Doppler shift to measure the velocity of plasma in the Sun's surface and solar atmosphere – important data for understanding the nature of solar eruptions.

A further insight provided by spectral lines occurs due to Zeeman splitting, which we touched on in Chapter 1. In the presence of a strong magnetic field, spectral lines will split into two, via the Zeeman splitting process. Zeeman splitting is a result of some complex atomic physics which we will not cover in this book. George Hale's observation of Zeeman splitting in sunspot spectra revealed the presence of strong magnetic fields on the Sun. Today, we use Zeeman splitting to measure magnetic fields over the Sun's entire surface, every minute of every day. The study of spectral lines under the presence of magnetic fields is known as *spectropolarimetry*. (Spectropolarimetry is the second longest technical word in this book, falling short

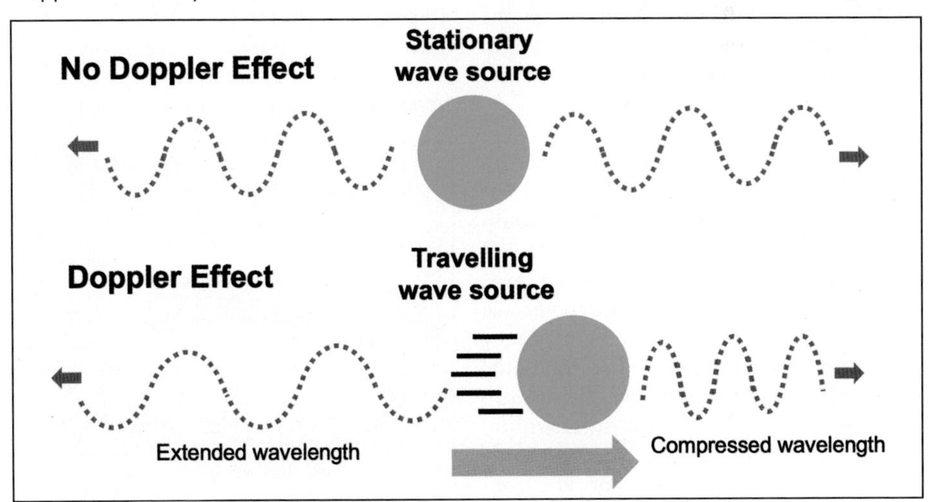

Doppler shift of a moving wave source.

of magnetohydrodynamics by two letters). Measurements of temperatures, densities, velocities, and magnetic fields on the Sun are critical for solar physicists to further understand the mysteries of our local star.

Magnetic Fields on the Sun

The simplest magnetic field is created by an isolated bar magnet, with one north pole and one south pole. This is called a *dipole*, with *di* meaning two. It is a universal rule in physics that the two poles of a magnetic field cannot be separated – magnetic monopoles cannot exist. If you were to cut a bar magnet in half, each half of the magnet would still have a north and south pole. If you've ever played with magnets, you'll remember that opposite polarities of a magnet (the north and south pole) attract, whilst two matching polarities repel each other. In a simple experiment using iron filings, you can see the larger influence of the invisible force field surrounding a magnet. This iron filings experiment is common in schools, done by simply placing a bar magnet in a tray of unordered iron filings. In the presence of the magnet, the iron filings will align themselves to the magnetic field lines to reveal the structure of the magnetic field. The magnetic field lines traced by the filings highlight the direction and strength of the magnetic field in the space around the magnet.

In an initial approximation, the Earth's magnetic field can be considered as a bar magnet, with one north pole and one south pole. The Earth's magnetic poles are not aligned with the Earth's rotational axis, which has its own north and south poles. We differentiate between these as the magnetic poles and rotational poles. The rotational north pole is constant, sitting at the very top of the globe. Magnetic north drifts around at a rate of about 55 km per year, currently located in northern Canada, 500 km from the rotational north pole. The magnetic poles can flip entirely, placing the magnetic north pole in the southern hemisphere. This last happened about 780,000 years ago.

The Sun's magnetic field is far more complex than the Earth's. This is primarily due to differential rotation, the varying rotation speed of different heights and latitudes across the Sun's surface. At solar minimum, when there are no sunspots on the Sun, the Sun's magnetic field is uncomplicated, and close to that of a dipole field. Because the Sun rotates with different speeds at different latitudes, the simple dipole field begins to 'wind up'. The winding up of the field adds complexity to the magnetic field. Eventually, regions of dense and complex magnetic field emerge from the Sun's surface, orientated with different polarities in each hemisphere. In the photosphere, the dense magnetic field repels the inflow of hot plasma, cooling down to become dark sunspots. Above sunspots, in the corona, the dense magnetic field lines fill with plasma to form bright *active regions* in the Sun's atmosphere. As the solar cycle continues and the global magnetic field is 'wound up' from fast-rotating high latitudes, the regions of complex field emerging at the surface migrate farther towards the equator. This explains the famous butterfly diagrams sketched by the Maunders. Eventually, as the wound-up field approaches the equator, it begins to cancel out with the magnetic field

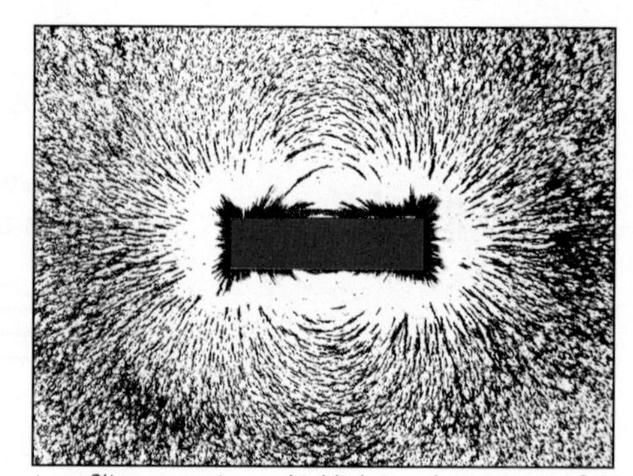

Iron filings experiment highlighting the magnetic field of a bar magnet.

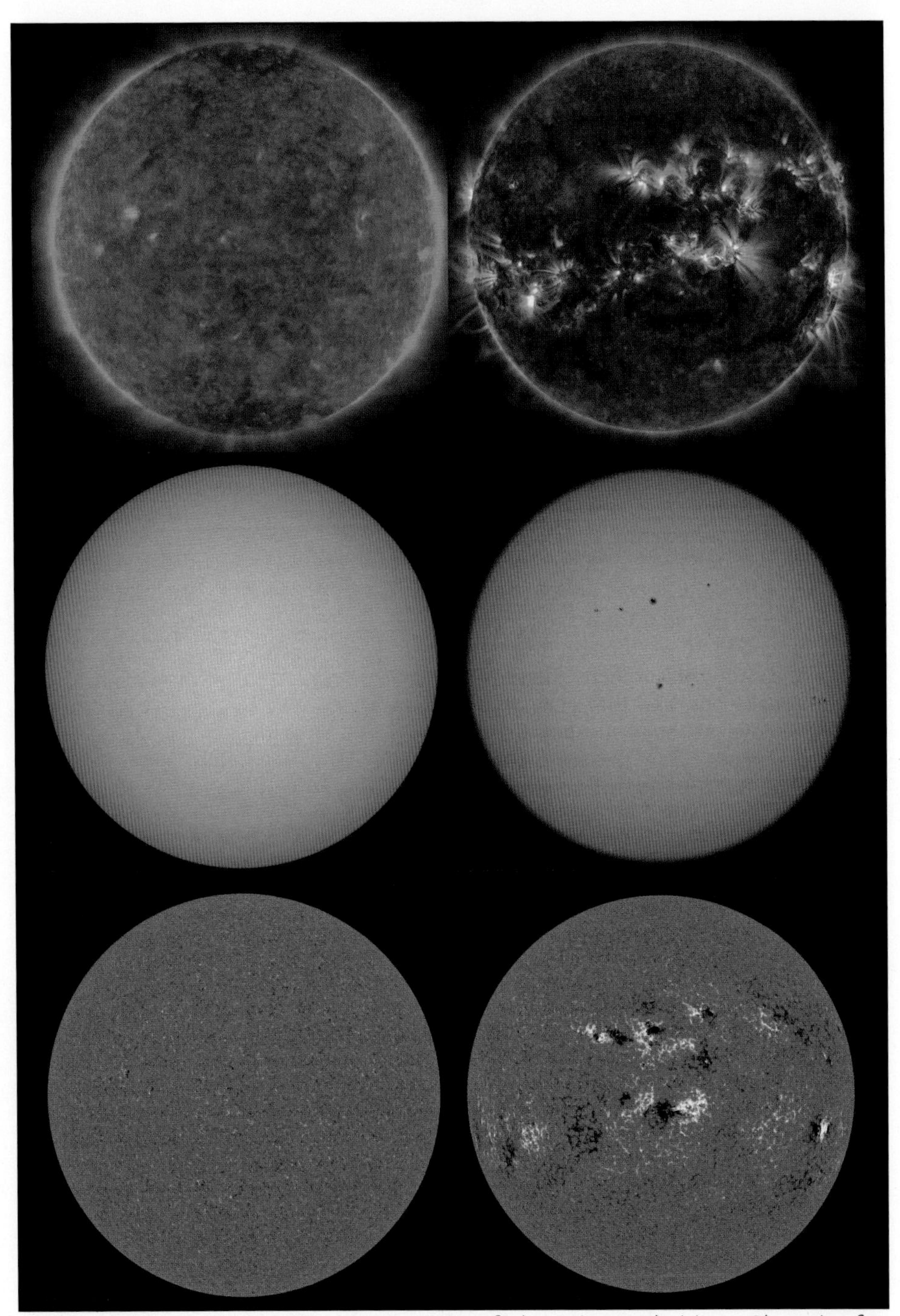

Corona, photosphere, and magnetogram image comparison of solar maximum and minimum. Observations from the NASA Solar Dynamics Observatory.

approaching the equator from the other hemisphere, which is the opposite polarity. As the complex magnetic fields from each hemisphere disappear from this interaction, the Sun is left with a simple dipole field once more, with an opposite polarity to what it began with. This entire cycle takes 11 years, driving the solar cycle between solar minimums with a simple global field to solar maximum at the peak of the magnetic complexity. The more complexity in the Sun's magnetic field, the more sunspots will emerge on the surface.

As George Hale discovered, each sunspot has its own strong magnetic field, with regions of north and south poles within them. With enough sunspots on the Sun, the photosphere is essentially littered with strong magnets. Interesting things happen in the active regions above sunspots. Just like iron filings around a bar magnet, solar plasma is bound to vectors of magnetic field in the Sun's atmosphere. These are called coronal loops, and they glow bright in active regions. Although sunspots on the surface are cooler than the surrounding photosphere, the opposite is true for active regions in the corona. As we will soon discuss, the strong magnetic field within these regions pumps energy into the system. This creates the perfect high energy conditions necessary for solar flares and coronal mass ejections to occur.

Using Zeeman splitting, we can reliably measure the magnetic field of the photosphere, producing what we call a magnetogram. The image on the previous page shows two magnetograms, one at solar minimum and one at solar maximum. The magnetograms are black and white in colour, showing areas of north and south magnetic field respectively. As you can see, the magnetic field is far more complex at solar maximum. Alongside the magnetograms, the images show simultaneous observations of the Sun's

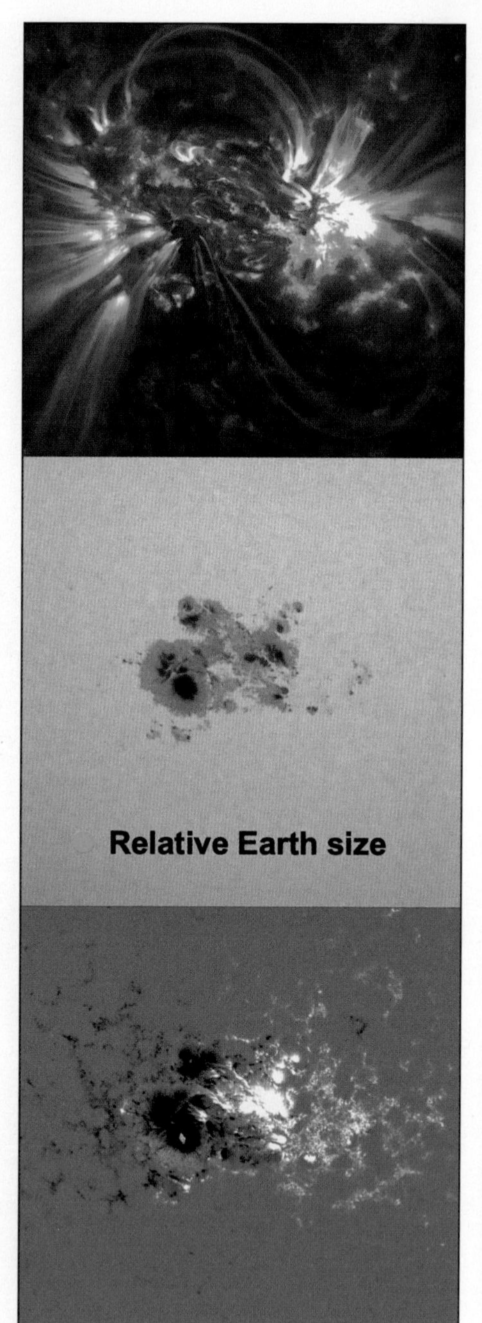

Corona, photosphere, and magnetogram image of a single active region. Observations from the NASA Solar Dynamics Observatory.

photosphere and corona, demonstrating the alignment of sunspots (in the photosphere) and magnetically complex active regions (in the corona). The image to the left shows similar observations, zoomed in for a single active region.

Solar Flares and Coronal Mass Ejections

Active regions are areas of complex magnetic fields in the Sun's atmosphere, often sitting above sunspots. Sunspots are constantly evolving, changing shape, and increasing or decreasing in magnetic field strength. As the sunspots change on the surface, the coronal loops above them, which are connected to the sunspot, are dragged around, twisted, and tangled. Opposite polarities of magnetic fields wish to repel one another, as with two simple bar magnets on a table. In active regions, field lines twisted and pulled together are not always able to move away from one another, which creates the build-up of something called magnetic free energy. To return to the bar magnet analogy, by pushing the north end of two magnets together, I am converting the chemical energy from my body (and the food I've consumed) into kinetic energy of the motion of my hands. This kinetic energy brings the two magnets together, which, as they wish to repel each other, builds free magnetic energy between the magnets. As I release the magnets, the free energy is converted back into kinetic energy of the magnets – they move to separate themselves. The same physics applies as I lift something off the ground. If you were to lift this book above your head, you'd convert the chemical energy of your body into kinetic energy of the book, creating gravitational potential energy. When you release the book, the potential energy is converted back into kinetic energy of the falling book. Both these scenarios are essentially storing energy for a time, either as gravitational potential or magnetic free energy.

This happens on the Sun. The physics is more complex, but you can use the previous examples as an approximation. In active regions, the twisting and tangling of the loops force the magnetic field into a configuration it would rather not be in, building free magnetic energy. As free energy continues to build up in the active region, the system wants to get rid of this excess energy. Eventually, the build-up of free energy will reach a breaking point.

Under normal circumstances in the solar corona, coronal magnetic loops, which trace vectors of magnetic fields, cannot pass through one another. If they could, the magnetic free energy would not build up to begin with. However, when the free energy becomes too large, conditions can arise where magnetic field lines can diffuse through one another. This diffusion region is called a current sheet, because the changing magnetic field across the region creates a large electrical current (think back to electromagnetism). Current sheets are incredibly thin, with a thickness of only 10 metres or so, tiny compared with the sizes of active regions on the Sun. Within this current sheet, magnetic fields behave differently from normal. Magnetic fields flow into the region and leave it with a completely different, lower energy configuration. This process, which removes magnetic free energy from the system, is called *magnetic reconnection.* Magnetic reconnection can completely reconfigure the entire magnetic field of the active

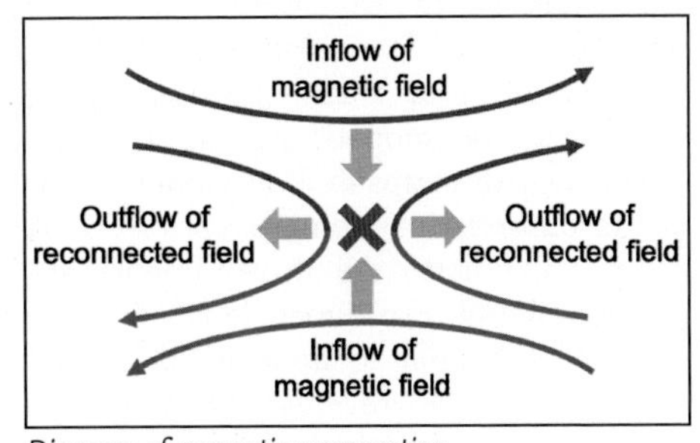

Diagram of magnetic reconnection.

region, producing a more relaxed magnetic topology.

The free energy released by magnetic reconnection cannot just disappear – it must go somewhere. In the solar corona, the released magnetic free energy is converted into several forms:

- **light**: from gamma waves to radio waves, energy is released all across the electromagnetic spectrum.
- **plasma heating**: the plasma, sitting along the newly reconnected field lines, is heated to incredible temperatures rising from around 1 million degrees to way over 10 million.
- **particle acceleration**: particles in the corona, such as protons and electrons, are accelerated to very high energies. Some of these are accelerated down to the surface of the Sun, whilst some are accelerated out into the solar system.
- **kinetic energy**: large magnetic structures can be untethered from the Sun via magnetic reconnection and accelerated to high velocities into the solar system.

This energy release has a name. The light, plasma heating, and particle acceleration created through the magnetic reconnection manifest themselves as what we call a *solar flare*.

The diagram to the right shows the standard model for eruptive solar flares, first developed by a few scientists in the 1960s–70s. The model holds up relatively well today and is capable of explaining many of the observed phenomena seen in solar flares. Solar flares are three-dimensional, but examining two-dimensional slices is conceptually easier to understand. In this model, solar flares are initiated as oppositely oriented magnetic field lines forced towards one another. Eventually, as magnetic free energy builds, magnetic reconnection begins, continuing for as long as field lines inflow into the current sheet region. The current sheet contains physics we do not yet fully understand, but out of the region flow newly reconnected field lines, above and below the current sheet.

The newly formed field lines beneath the current sheet are still attached to the Sun's surface, forming relaxed loop structures called flare loops. High-energy particles are accelerated from the reconnection site down the flare loops, where they collide into the Sun's surface below. They impact with the Sun's chromosphere, a thin layer between the photosphere and base of the corona. The high-energy particles smash into the chromosphere, causing the cooler, dense chromospheric plasma to heat up. This causes the intersection between the flare loops and chromosphere to glow brightly in ultraviolet and hard (higher energy) X-rays, forming flare ribbons that move and grow as the flare loops evolve. As the chromospheric plasma heats up, it rises to fill the flare loops with hot material, causing the flare loops to glow brightly in ultraviolet and soft (low energy) X-ray observations. The flare loops eventually cool down, and new hot loops continue to form above them for the lifetime of the solar flare, which typically lasts for tens of minutes. Both flare loops and flare ribbons are frequently observed by space-based telescopes. In some cases, they can even be spotted by amateur telescope setups too, but this requires a bit more luck with the timing. It was these bright flare loops that Richard Carrington observed in a sunspot before the Carrington Event in 1869 – he was the first human to witness a solar flare.

Returning to the diagram of the standard model of solar flares, newly reconnected field lines are also expelled above the

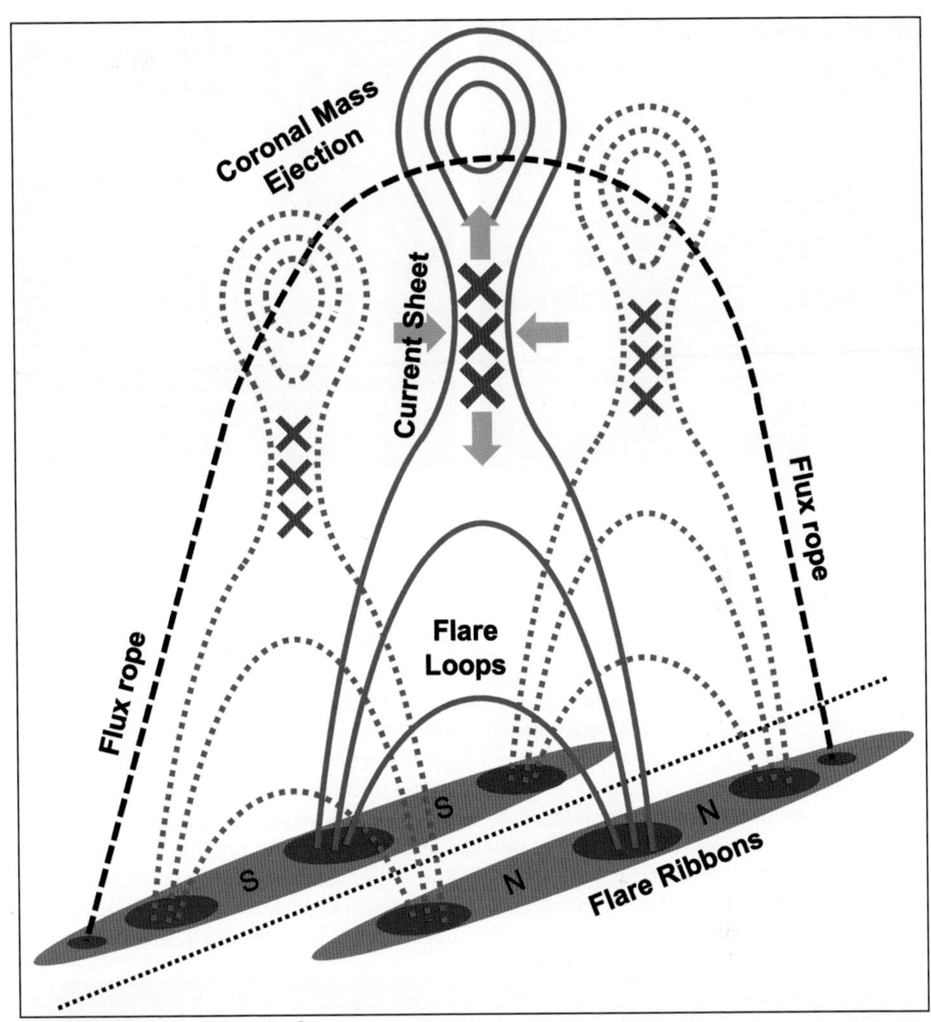

Standard model for eruptive solar flares.

current sheet. Unlike the field lines beneath the current sheet, the field lines above are no longer anchored to the Sun. In three dimensions, this region is not completely untethered, but connected to the surface by a complex magnetic structure sitting inside the newly reconnected loops. This structure is called a *flux rope*. Magnetic reconnection weakens the connection between the flux rope and the Sun's surface, cutting its ties to the Sun by reconfiguring the field lines initially pinning it down. This causes the flux rope to rise slightly, which tightens the magnetic field lines around it, forcing additional field lines into the current sheet and further feeding magnetic reconnection. This causes a positive feedback mechanism, as continued reconnection accelerates the rising flux rope, which in turn creates even more reconnection. This results in a solar eruption, as the flux rope and surrounding magnetic field accelerate away from the Sun, reaching speeds up to 3,000 km/s. By solar system standards, 3,000 km/s is fast,

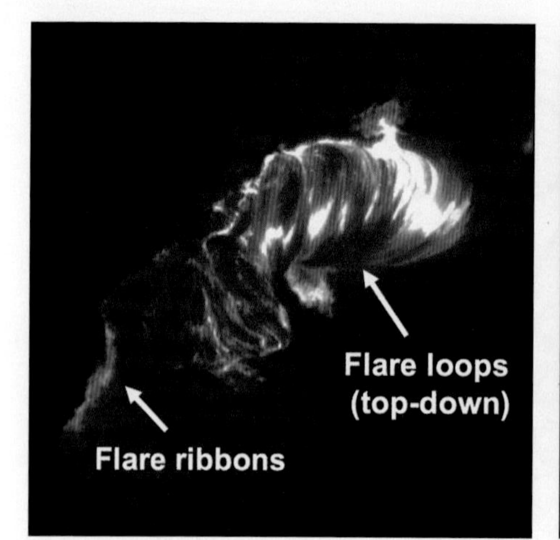

Flare loops and flare ribbons, captured by SDO/AIA 304 Å.

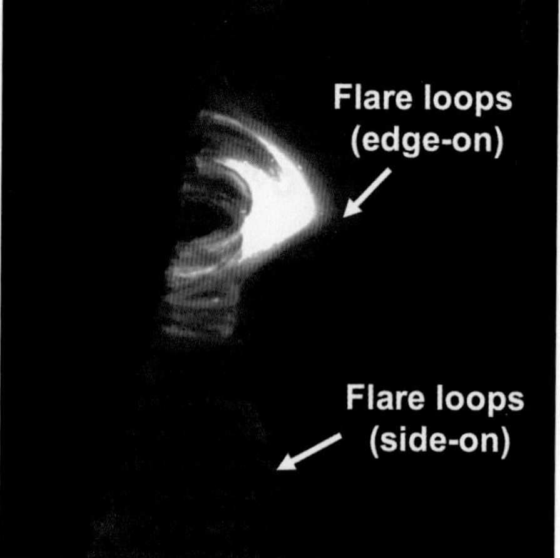

Edge-on view of a flare loop arcade, captured by SDO/AIA 304 Å.

and solar eruptions traveling at this speed can span the distance between the Sun and Earth in less than a day. This eruption is known as a coronal mass ejection, or CME for short. CMEs, although unknowingly observed during the 1860 eclipse, were not discovered until adequate space-based observations became available in 1971.

CMEs can erupt at any angle relative to the Earth. If we see them erupt edge-on, we get a clear view of their direction, speed, and structure. If they erupt directly towards us, we instead see a CME halo as it expands outwards whilst travelling in our direction. This is called a *halo CME*. Equivalently, if the eruption is on the far side of the Sun, erupting away from us, we will also see a halo CME. In recent years, attention has turned to a special variety of CMEs called a *stealth CME*. A stealth CME occurs when a halo CME is clearly visible, but with no apparent source on the Sun. Solar physicists used to attribute such CMEs to eruptions on the far side of the Sun. The problem was, on some occasions, we'd detect the CME arriving at Earth a few days later – so it is clearly Earth-directed.

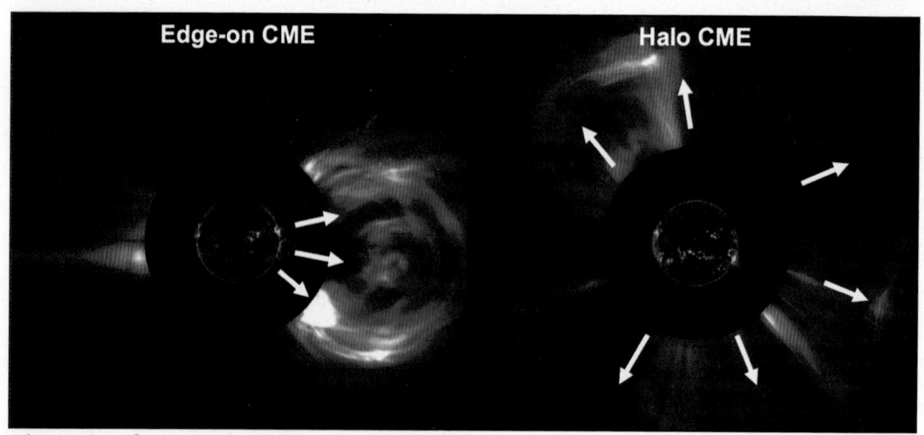

Edge-on view of a CME and a head-on view (a halo CME)

Why such eruptions exist and why we don't see signatures of them in the low solar atmosphere, is an area of ongoing research. It also stresses the importance of having telescopes situated beyond orbit around the Earth, to provide a different vantage point of CMEs travelling in the Sun-to-Earth direction. Such a spacecraft would tell us for certain whether and when an eruption is heading directly towards us, and enable us to monitor the CME evolution as it travelled towards the Earth.

The standard model of eruptive solar flares is just one mechanism through which solar flares and coronal mass ejections can occur. Solar flares and CMEs can even occur completely independently of one another, and only correlate in about 50% of instances. There are several other models of explosive events on the Sun, some producing flares without an eruption and some producing CMEs without significant magnetic reconnection. Stealth CMEs are likely an example of this latter scenario.

Solar flares are categorised by their soft X-ray output. There are four categories of solar flares, each ten times more powerful than the category before. These categories are B-, C-, M-, and X-class flares. Although the main process in solar flares, magnetic reconnection takes place all over the Sun, not only in high energy explosive events. When the energy released through the reconnection is large enough, we observe the process as a solar flare. Small-scale brightenings and jets are present all over the Sun too, and whenever we build a new telescope to examine the Sun in more detail, we see brightenings on even smaller scales still. We also predict that very small reconnection events, called *nanoflares*, occur on scales we'll never observe. We suspect that these tiny events must exist in order to explain why the Sun's atmosphere is so much hotter than its surface. More on this later.

Solar Wind

The space between the Sun and the Earth is not empty, but filled with a constant stream of low-density plasma emanating out of the solar atmosphere. This stream of plasma, called the *solar wind*, travels at an average speed of 400 km/s, with a density of under 10 particles per cubic cm – about 10 billion billion times less dense than Earth air at sea level. Like plasma in the solar corona, solar wind plasma is also bound to magnetic fields, and drags the magnetic field with it as it emanates away from the Sun. This magnetic field, which extends way past the orbit of Pluto, is known as the *heliosphere*. Seven of the planets in our solar system (Mercury, Mars, Earth, Jupiter, Saturn, Uranus, and Neptune) have their own planetary magnetic fields (although those of Mercury and Mars are very weak), and these exist as magnetic 'bubbles' within the much larger magnetic field of the Sun. The magnetic field of a planet is called its *magnetosphere*.

The Earth's magnetic field can be approximated as a dipole (a simple magnet with one north pole and south pole) close to the Earth, but it is heavily distorted by the solar wind at higher altitudes. At the dayside of the Earth, where Earth faces the Sun, the solar wind impacts the Earth's magnetic field and compresses it. The sudden deceleration of the solar wind forms a *bow shock* (like the shock at the bow of a ship as it moves through water), behind which the Earth's *magnetopause* defines the sharp boundary of the Earth's magnetosphere. At the nightside of the Earth, the Earth's magnetic field is stretched downstream of the solar wind, forming the Earth's *magnetotail*. Returning to the ship analogy, the magnetotail is equivalent to the wake of the ship.

The heliosphere is the full extent of the Sun's magnetic field. Some astronomers use the heliosphere boundary to define the edge of the solar system, although other definitions do exist. As the Sun moves through the galaxy, dragging its magnetic field with it, it creates a similar structure to the Earth's magnetosphere. The front edge of the heliosphere is called the *heliopause,* which sits behind the bow shock in interstellar space. Behind the Sun, there is a long magnetotail stretching out many times farther than the distance to the heliopause. In 1977, NASA launched two spacecraft, *Voyager 1* and *Voyager 2*, into the solar system. The scope of the mission was for the spacecraft to conduct flybys of the outer planets, measuring the environments they flew through and capturing photos of the outer planets for the first time. The mission was a huge success. The close-up images of Jupiter and Saturn were the best images we'd see for decades until dedicated missions to those planets took place in the 1990s. For Uranus and Neptune, we haven't got closer since. After achieving the primary goal of the mission, the two Voyager spacecraft were left hurtling towards the edge of the solar system at speeds of 61,500 km/h, measuring the space environment as they flew through it.

In the decades that followed, *Voyager 1* continued measuring the density and magnetic field of the path it flew through, beaming the data back to Earth. But all of a sudden, in August 2012, something changed. The readings on *Voyager 1* looked strange, and after some analysis, scientists confirmed that the spacecraft had crossed the heliopause, leaving the heliosphere and entering interstellar space. It took 35 years and 18 billion km (122 times farther from the Sun than the Earth) for it to reach this point. Six years later, *Voyager 2* achieved the same milestone. The two spacecraft remain the only human-made objects to leave the heliosphere and won't be joined by another spacecraft until after 2030, when the *New Horizons* spacecraft (which passed Pluto in 2015) crosses the boundary too. To this day, both *Voyager 1* and 2 continue to fly into the abyss of interstellar space (and will continue to do so for many many more years to come).

The solar wind speed is not constant, but is determined by its source region on the Sun. The quiet Sun produces an average speed of 400 km/s, but large holes in the solar atmosphere, called *coronal holes*, can create solar wind speeds of up to 700 km/s. When a coronal mass ejection erupts from the Sun, the CME joins the solar wind, travelling at speeds of 2,000 km/s (or even faster). Like a snow plough, a fast-moving CME will scoop up the slower solar wind in front of

Artist's impression of the Earth's magnetosphere in the solar wind.

Jupiter

Neptune

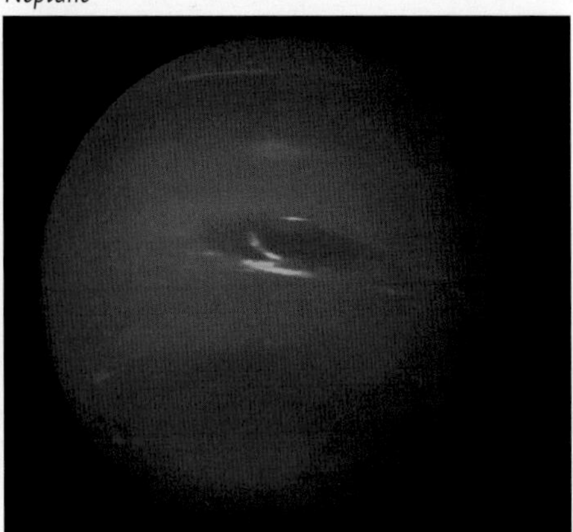

Saturn

Uranus

Voyager 1 and 2 photos of Jupiter, Saturn, Uranus and Neptune, as they flew through the heliosphere.

it, increasing its density as it travels through the solar system. The magnetic structure of an erupting CME is maintained in the solar wind and measured by spacecraft 'tasting' the environment around them. Some of these spacecraft are sitting 1.5 million km from the Earth, about 1% of the distance to the Sun. Other spacecraft, such as the NASA *Parker Solar Probe* or ESA *Solar Orbiter* orbit closer to the Sun, measuring the solar wind at distances much closer to the Sun than the Earth is. Although the solar wind always travels radially away from the Sun, due to the Sun's rotation, the magnetic field of the solar wind follows a spiral shape called the *Parker spiral*. To visualise this, imagine yourself spinning around whilst pointing a garden hose outwards. Each drop of water shoots straight away from you, but due to the spinning water source (you), a spiral is visible in the overall system. Because of the Parker spiral, the solar wind and eruptions from the very centre of the Sun (from our perspective) will not hit us. Instead, the Parker spiral will carry the plasma thousands of kilometres to our left (when looking at the Sun with the Sun's north pointing up), missing us completely. The solar wind arriving at Earth originates from the middle-right of the Sun, which makes eruptions from this location the most dangerous.

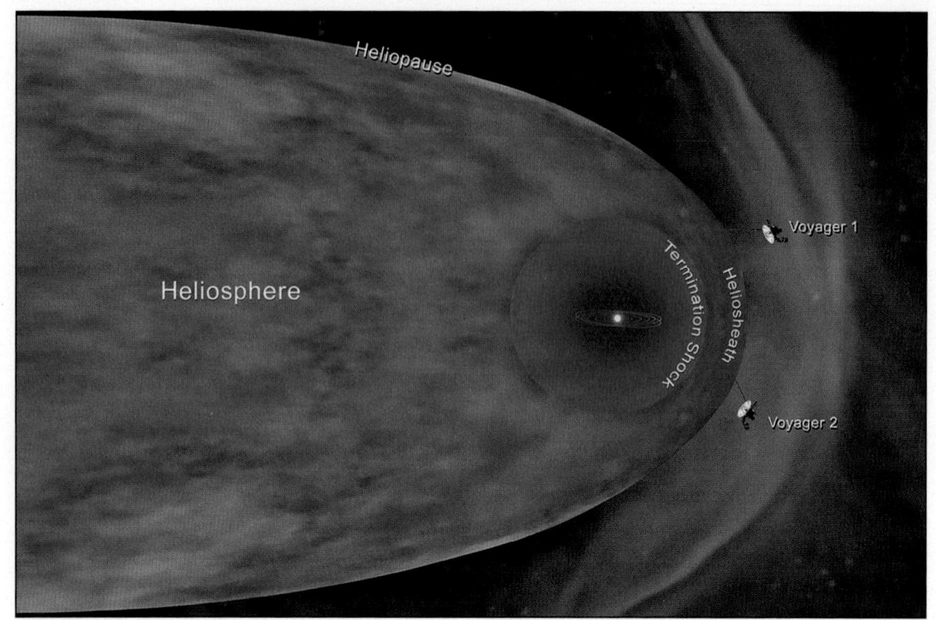

The heliosphere and Voyager spacecrafts (not to scale).

Aurora Borealis and Aurora Australis

The most impressive aurora on human record occurred during the Carrington Event of September 1859, which we covered in Chapter 1. Earlier that day, Richard Carrington observed a flash of light on the Sun, which we now know was a solar flare. Even more bizarre than the intense aurora was the malfunction of telegraph machines across Europe, which continued to send and receive messages without any power. What really happened on that day?

The solar flare observed by Carrington was a big one. Solar flares commonly cause brightenings in ultraviolet light, but only the largest flares produce white light bright enough to see against the bright background Sun. The flare would have triggered a large coronal mass ejection (CME), which from Carrington's observations we know was positioned perfectly on the Sun to follow the Parker spiral directly towards Earth. Because of the rotation of the Sun, if the eruption had happened a few days earlier or later, it would not have been Earth-directed. When the complex magnetic field of the CME impacts the Earth's magnetosphere, some interesting physics happens to lead to the formation of the aurora.

Magnetic reconnection, the main driver of solar flares, is also a key mechanism in creating the aurora. Because magnetic reconnection only occurs between oppositely oriented magnetic fields, the orientation of the CME upon impact with the Earth's magnetosphere is important. The orientation of the Earth's magnetosphere does not change (not over the timescale of these processes at least), with a consistent orientation of magnetic north pointing 'upwards'. A CME, on the other hand, can have a magnetic field pointing in any direction, depending on the magnetic configuration of its source active region and journey from the Sun to the Earth.

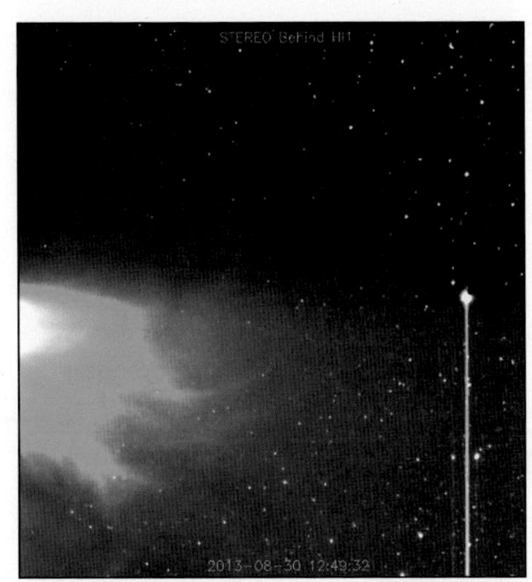

Solar wind observed by heliospheric imager onboard the NASA STEREO-B spacecraft. The Earth is the bright dot on the right, and the Sun is out of frame to the left.

Solar wind observed by the WHISPR instrument on the NASA Parker Solar Probe.

If the incoming CME also has a northern-orientated magnetic field, then magnetic reconnection cannot occur. In this instance, the CME will simply flow past the Earth, like a stream around a big rock. When the opposite is true and the incoming CME has a southern-orientated magnetic field, we get a more interesting scenario. At the dayside of Earth, where the CME impacts, magnetic reconnection takes place between the CME and Earth's magnetosphere to create a direct magnetic connection between the solar wind and Earth's poles. This open connectivity cannot last for long, as, due to the fast-flowing solar wind, the newly reconnected field lines are dragged downstream of the Earth. The magnetic field lines are stretched out either side of the magnetotail at the nightside of the Earth. Because these field lines are opposite in polarity, magnetic reconnection occurs once again – this time in the magnetotail. The magnetic reconnection process closes the Earth's magnetosphere again, but pumps the newly reconnected field lines with high energy particles. Some of these particles come from the CME directly, but most are accelerated by energy release during magnetic reconnection. (This process of dayside to nightside reconnection is named the *Dungey cycle*, after the astronomer James Dungey who first proposed the concept in 1961).

The high-energy particles follow the Earth's magnetic field down towards the Earth's poles, where they collide with cooler gases in the Earth's atmosphere. The collisions from high-energy particles excite electrons in the atoms of the Earth's atmosphere. Just like on the Sun, the atoms will eventually dispose of this unneeded energy, converting the energy into light (at the corresponding wavelength to the electron's energy). The glowing of the atmosphere as a result of this process is how the aurora is created. Essentially, the aurora is the emission spectrum of the Earth's atmosphere, with the wavelengths of light emitted acting as signatures of the elements present in the upper atmosphere. This is why the aurora has colour. The classic green colour is created by electrons within an oxygen atom, and the rarer blue and red colours from nitrogen.

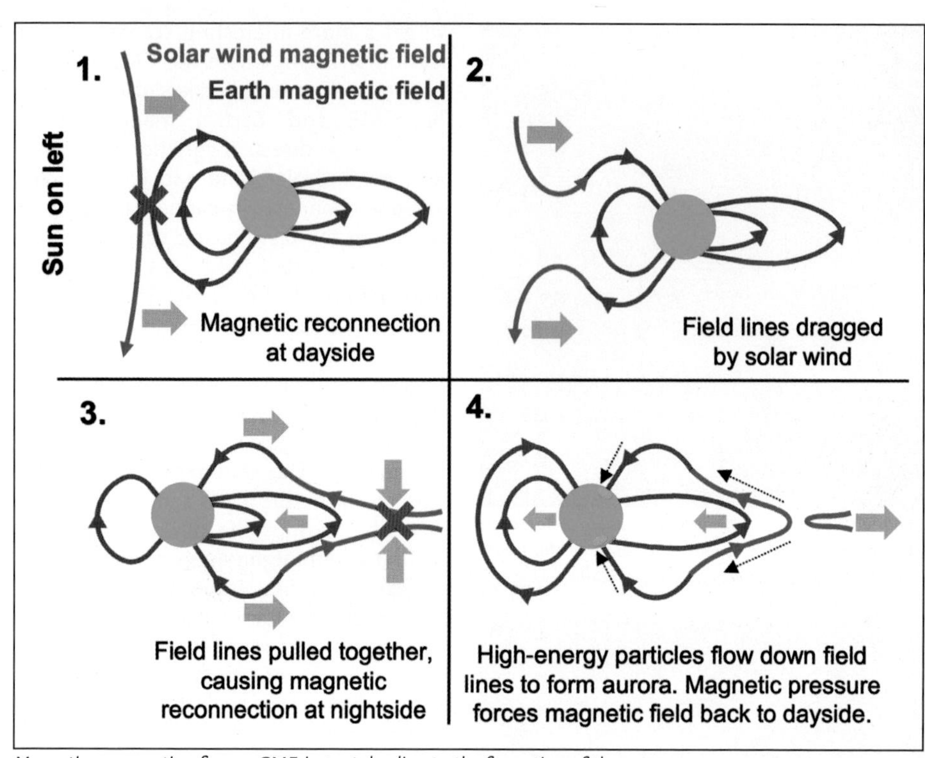

Magnetic reconnection from a CME impact, leading to the formation of the aurora.

The stronger the southward component of the CME, the more layers of the daytime magnetosphere it can 'peel off' through magnetic reconnection. Although the size and speed of the CME are important, its orientation is more so. As more and more field lines reconnect, high-energy particles intersect with the Earth's atmosphere progressively farther from the poles. During the Carrington Event of 1859, the southward component was so strong that the aurora was visible all the way down to the tropics. Ordinarily, the aurora is reserved for regions much closer to the poles.

The disturbance of the Earth's magnetic field from the CME impact is called a geomagnetic storm, and the ability of a CME or solar wind stream to cause a geomagnetic storm is known as its *geoeffectiveness*. When a CME erupts from the Sun, we have no idea how geoeffective it will be until it has nearly arrived at Earth. Minor geomagnetic storms can also be caused by the fast solar wind stream of a coronal hole, but the largest geomagnetic storms are nearly always caused by CMEs.

On page 56 are photos of the Aurora Borealis, a wonderful sight if you ever get the chance. The top photo was taken by a professional aurora chaser, and the bottom one by an aurora first-timer with their phone. If you want to learn more about the aurora from the Earth's perspective – it's history, science, and photography tips – there is a dedicated book for that in the Collins Astronomy collection, *Northern Lights: The definitive guide to auroras*, written by my mentor and good friend Tom Kerss.

Different colours of the aurora.

The northern lights photographed by a professional aurora chaser, Vincent Ledvina.

The Northern Lights photographed by an aurora first-timer, David Wildgoose.

Space Weather in the Modern Age

The impacts of a geomagnetic storm extend further than the aurora, and they also have the ability to affect our satellites, power grids, and other technological infrastructure. The influence of solar activity on our technology is called space weather. Space weather consists of three main categories, geomagnetic storms, *solar radiation storms*, and *radio blackouts*, each relating to a different event on the Sun.

- Geomagnetic storms are caused by magnetic reconnection in the magnetosphere (resulting from a CME or fast solar wind stream).
- Solar radiation storms are caused by the impact of high energy particles accelerated in a solar flare.
- Radio blackouts are produced by the reaction of the Earth's upper atmosphere to the intense emission of light (mainly X-rays) from a solar flare.

The next sections introduce the three categories of space weather in more detail.

Geomagnetic Storms

Geomagnetic storms are the variation of Earth's magnetic field under the influence of solar activity. They are caused by magnetic reconnection at the dayside and nightside of the magnetosphere following the impact of a southern-directed CME or fast solar wind stream. CMEs create larger geomagnetic storms than fast solar wind streams but are much harder to predict. Fast solar wind streams originate from coronal holes, regions of open magnetic field in the Sun's atmosphere (literally holes in the corona). Because coronal holes are fairly long-lived, we are able to predict by around a week in advance when a fast solar wind stream will become geoeffective.

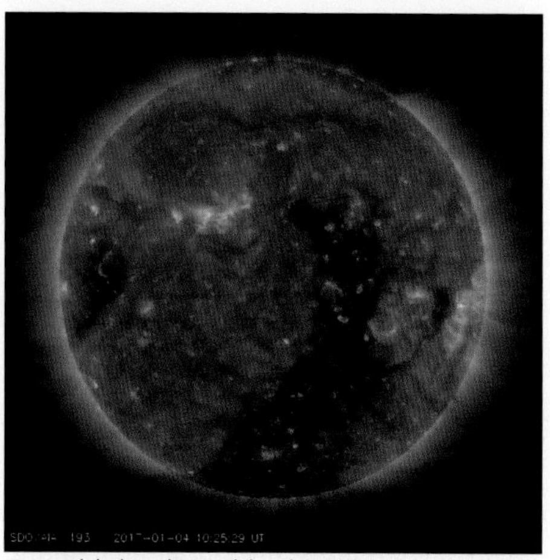

Coronal holes observed by the NASA Solar Dynamics Observatory.

Magnetic and electric fields are directly related through electromagnetism. Varying a magnetic field will produce an electric field, which can create an electrical current within a conductive object. When the Earth's magnetic field fluctuates during a geomagnetic storm, the process induces a current along distances of long conductive materials. Unfortunately for us, our power lines, telegraph wires, and railway networks, as is their design, are built of exactly these. During the 1859 Carrington Event, the electric field generated along telegraph wires was strong enough to send messages between telegraph machines even with no external power source. For telegraph machines in operation, the addition of an unwanted electric current to normal electricity in the machines caused them to overload, creating sparks and shocking operators. In general, a fast-varying electric field is unkind to electrical components not designed to withstand them. In our houses, rapid flicking of a switch or the use of a faulty device will trip a fuse. This is by design, to avoid permanent damage to the system. The electric current induced by a geomagnetic storm is much larger than that from a faulty hairdryer, and unfortunately, nowadays there is more to worry about than telegraph wires.

In space, satellites sitting within the varying magnetic field will experience an induced electric current beyond what they were designed to tolerate. This has the potential to temporarily or permanently damage the satellites in orbit, on which we depend for everyday life. On the ground, railway lines built of miles of straight metal rods (a perfect electrical conductor) could similarly experience technical faults. Elsewhere, power grids are at risk. Power lines and transformers carry an electric current of their own, so the additional electric field induced by a geomagnetic storm can damage these components. In the case of a damaged transformer, localised regions could be without power for hours or even days. In 1989, power grids across the east coast of Canada collapsed for 13 hours due to a CME impact, leaving millions of people without electricity.

Like the aurora, negative effects of a geomagnetic storm are felt most strongly near the poles, descending to lower latitudes as the strength of the geomagnetic storm increases. The strong widespread impact of the Carrington Event is often considered as the worst-case scenario for geomagnetic storms – the worst scenario that space weather forecasters (we'll talk about them later) can hope to prepare for. In the case of a Carrington Event, up to 15% of satellites could be permanently damaged and localised regions subjected to power blackouts for several days. The economic impact of this would rival the cost of the worst natural disasters.

Solar Radiation Storms

Solar flares produce light, heating of plasma, and acceleration of high energy-particles. Accelerated protons and electrons from solar flares can move either towards the Sun or out into the solar system. Particles travelling in the latter direction are known as *solar energetic particles*, which produce solar radiation storms here on Earth. Solar energetic particles travel only slightly slower than the speed of light, arriving at Earth minutes after the solar flare is observed. Components in computer systems send signals to one another by firing electrons. Inside a computer motherboard, electrons fire all over the system, activating different complex processes. Most solar energetic particles cannot penetrate the Earth's atmosphere, but computer systems on board satellites are at risk. The onboard electronic systems cannot tell the difference between an electron fired by another component on board and an electron from the Sun. Rogue electrons from the Sun produce *single event upsets* in satellite computer systems by activating random functions which can even render the satellite unusable. Most single event upsets can be fixed by a 'power cycle' (a fancy term for turning it off and on again). However, restarting a satellite contains a lot of risk, as they are not often tested for this scenario, so have a chance of never working again. A solar radiation storm is a shower of these solar energetic particles from the Sun.

On very rare occasions, solar energetic particles can penetrate through the atmosphere to produce a 'Ground

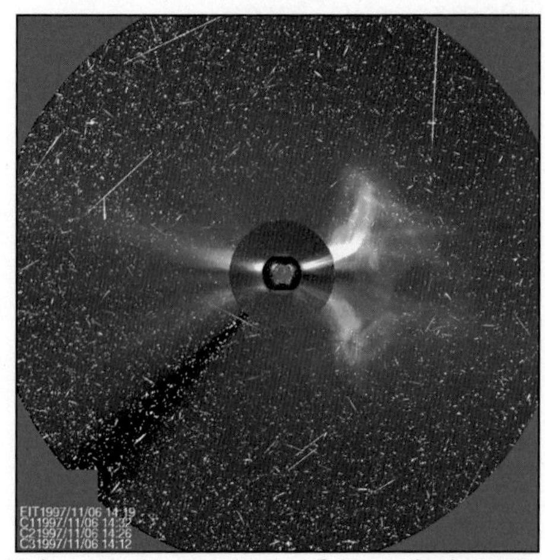

Solar energetic particles interfering with the NASA LASCO instrument as it observed a solar flare.

Level Enhancement'. These events are comparatively rare, with only a few on record. Solar radiation storms can produce radiation levels harmful to humans. Astronauts are particularly at risk and have procedures setting out where to seek refuge on the International Space Station if such conditions are forecast. On Earth, pilots, passengers, and crew flying at high latitudes are also exposed to an increase in radiation levels. This is not instantaneously dangerous, but repeated exposure to many events can increase the risks of diseases such as cancer. Thankfully, large solar radiation storms do not occur with enough frequency to cause any significant risk to the aviation community.

Radio Blackout

The final category of space weather is called a radio blackout. Radio blackouts occur as intense X-rays produced by a solar flare hit the upper atmosphere. X-rays are harmful to humans, and exposure to them should be kept as low as possible over a human lifetime. Thankfully the Earth's atmosphere absorbs X-rays, and so no X-ray emission from a solar flare will reach us on the ground. However, as a result of absorbing X-rays, the upper layer of the atmosphere heats and expands. The heating of the upper atmosphere does not affect temperatures on the ground, but does impact our communication systems. Radio communication, which we use to communicate with satellites, aeroplanes, and ships, works by bouncing radio waves off the upper atmosphere, transmitting signals between different locations. During a solar flare, if the Earth's upper atmosphere expands due to the impact of X-rays, then radio waves are no longer able to propagate effectively. This effect is called a radio blackout and is greatest on the dayside of Earth. During a radio blackout, technology relying on the propagation of radio waves will no longer work as usual, and sometimes cease to work at all. Understanding when a radio blackout is likely is critical for planning disaster relief and military operations. In September 2017, a series of large flares occurred concurrently with Hurricane Irma, which hit the Caribbean hard. During the hurricane, rescue operations were hindered due to radio blackouts from the flares. In military situations, the consequences are even more serious, and it is crucial to distinguish between a loss of communication due to space weather and a targeted enemy attack. During the height of the Cuban Missile Crisis in 1967, US tracking systems lost communication due to a solar radiation blackout. The initial reaction was that this was the prelude to an enemy attack. Thankfully, before retaliation strikes were ordered, scientists realised that this was the work of the Sun, not an enemy state. Because of how badly this event could have gone, this was a turning point in the US for the funding of space weather research.

Global navigation satellite systems (GNSS) such as the United States-operated Global Positioning System (GPS) also rely on the propagation of radio waves through the atmosphere. During a large space weather event, signals from these satellites will be blocked from reaching the ground. The absence of GPS may not seem like a big deal at first, but many sectors, including aviation, shipping, emergency services, large-scale automated farming, and mining, rely on GPS to operate. Large-scale outages are not common, and we have yet to experience a large-scale blackout of satellite navigation since the technology's invention. Smaller-scale radio blackouts are frequent, though, and very relevant to specialised users reliant on radio technology.

UK National Risk Register

In 2010, the volcano Eyjafjallajökull erupted in Iceland. Winds carried the large

	Less than 0.2%	0.2 to 1%	1 to 5%	5 to 25%
Level E (catastrophic)	Civil nuclear accident			Pandemics
Level D (significant)		Reservoir/ dam collapse	Severe coastal or river flooding	**Severe space weather**
Level C (moderate)		Extreme droughts	Poor air quality	Extreme storms Volcanic eruptions
Level B (limited)		Wildfires Major fires	Widespread public disorder	
Level A (minor)	Earthquakes			

Summary of the UK National Risk Register. The table organises potential disasters by their probability of impacting the UK each year, and potential impact level, ranked from A (lowest) to E (highest).

plume of volcanic ash across northern Europe, which resulted in the closure of airspace for 8 days in April that year. This disruption was completely unprecedented and unexpected, costing the global airline industry $1.7 billion. Following these events, the United Kingdom and other governments around the world started to consider what other natural disasters they were unprepared for and what the outcome of these disasters might be. Shortly after this, in 2011, space weather was finally added to the UK National Risk Register. The National Risk Register is a list of all potential events threatening to impact the country negatively, both natural and human-caused. It is a long document, overviewing details for a long list of potential disasters. The table above, adapted from an official table featured in the UK National Risk Register, summarises the long list and demonstrates the relative threat of space weather alongside other disasters that might possibly impact the UK. The horizontal axis of the table shows the probability (as a percentage) of a disaster occurring each year, and the vertical axis gives the potential impact categorised from A (lowest impact) to E (highest impact). Note that the impact scale refers only to the impact on the UK, so does not reflect any wider geographical reach of the effects on the rest of the globe (which a severe space weather event would have, but regionalised flooding would not).

Forecasting Space Weather

There are groups around the world dedicated to forecasting space weather 24 hours a day, 7 days a week. The two primary non-military space weather forecasters are the Space Weather Prediction Center (SWPC) within the US National Oceanic and Atmospheric Administration (NOAA) and the UK Met Office Space Weather Operations Centre. Like regular weather (or as space scientists call it, terrestrial weather) forecasters, space weather forecasters produce twice-daily forecasts for space weather activity. Because we do not know when a solar flare will happen, producing forecasts is tricky. Space weather forecasters examine the magnetic complexity of active regions and publish a probability of a flare erupting. Once a solar flare has happened and radio blackouts and solar radiation storms begin, forecasters will also provide updates on the expected duration of the events.

Forecasting geomagnetic storms is slightly different. The geoeffectiveness of a coronal hole is easier to predict, but the large geomagnetic storms created by CMEs are more challenging. Forecasters will detect the eruption from the Sun, but this is followed by a delay of 16-36 hours before the CME arrives at Earth. They combine observations of the Sun with cutting-edge models to try to predict the speed, timing, and potential impact of the CME.

Space weather forecasters define the potential impacts of space weather with a 5-tier scale for each category of space weather. The Carrington Event, which is predicted to occur with a frequency of every 100-150 years, is often considered the worst-case scenario on these space weather scales. An overview of the space weather scales is as follows and is created by combining information from both the NOAA and Met Office forecasts:

Ancient Solar Flares

Although the Carrington Event is often considered the 'worst-case scenario' for space weather mitigation, we know that even

Geomagnetic Storms		
Scale	**Effect**	**Average frequency**
G5 - extreme (Kp = 9)	*Power systems:* Widespread collapse or blackouts in power systems, damaged transformers. *Spacecraft:* Corrective actions to satellite orientation and position may be required. Problems tracking and uplinking/downlinking to satellites. Extensive surface charging. *Satellite navigation:* Degraded for days. *Aurora:* Visible throughout the UK and down to Florida and southern Texas in the US.	4 days per cycle
G4 - severe (Kp = 8)	*Power systems:* Voltage control problems possible, but no significant impact. *Spacecraft:* Corrective actions to satellite orientation and position may be required. Potential surface charging. *Satellite navigation:* Degraded for hours. *Aurora:* Visible throughout the UK and down to Alabama and northern California in the US.	60 days per cycle
G3 - strong (Kp = 7)	*Power systems:* Voltage control problems possible, but no significant impact. *Spacecraft:* Corrective actions to satellite orientation and position may be required. Potential surface charging. *Satellite navigation:* Intermittent. *Aurora:* Visible down to Northern Ireland, mid-Wales, and the Midlands in the UK and down to Illinois and Oregon in the US.	130 days per cycle
G2 - moderate (Kp = 6)	*Power systems:* Voltage control problems at high latitudes possible, but no significant impact. *Spacecraft:* Corrective actions to satellite orientation and position may be required. *Aurora:* Visible across Scotland in the UK and down to New York and Idaho in the US.	360 days per cycle
G1 - minor (Kp = 5)	*Power systems:* No significant impact. *Spacecraft:* Minor impact on satellite operations possible. *Aurora:* Visible down to northern Scotland in the UK and down to Michigan and Maine in the US.	900 days per cycle

Table of geomagnetic storm effects. The G scale is a scale of the effects of the geomagnetic storm, whilst the Kp index is a physical measurement of the magnitude of the Earth's magnetic field variation due to the geomagnetic storm. (Kp index stands for 'planetary K index', where 'K' refers to the horizontal component of the magnetic field). The average frequency is per solar cycle (lasting around 11 years).

Solar Radiation Storms		
Scale	**Effect**	**Average frequency**
S5 - extreme	*Human health:* Unavoidable high radiation hazard to astronauts and increased exposure for passengers/crew of aircraft at high altitudes and latitudes. *Satellites:* Loss of satellite memory, imaging, tracking, and control. Permanent damage to satellite solar panels.	1 day or less per cycle
S4 - severe	*Human health:* Unavoidable high radiation hazard to astronauts and increased exposure for passengers/crew of aircraft at high altitudes and latitudes. *Satellites:* Problems with satellite memory, imaging, and tracking, degraded satellite solar panel efficiency.	3 days per cycle
S3 - strong	*Human health:* Radiation hazard to astronauts, and increased exposure to passengers/crew of aircraft at high altitudes and latitudes. *Satellites:* Satellite single event upsets, imaging noise, and slight reduction in satellite solar panel efficiency.	10 days per cycle
S2 - moderate	*Human health:* Increased exposure for passengers/crew of aircraft at high altitudes and latitudes. *Satellites:* Single event upsets.	25 days per cycle
S1 - minor	No significant impact on human health or satellites.	50 days per cycle

Table of solar radiation storm effects. The average frequency is per solar cycle (lasting around 11 years).

Radio Blackouts		
Scale	**Effect**	**Average frequency**
R5 - extreme (X20 flare)	Complete radio communication blackout, radio navigation blackout, and satellite navigation errors on the Earth dayside for many hours. (No communication or navigation for planes or ships.)	1 day or less per cycle
R4 - severe (X10 flare)	Radio communication blackout, radio navigation outages, and minor satellite navigation errors on the Earth dayside for 1–2 hours.	8 days per cycle
R3 - strong (X1 flare)	Radio communication blackout, and radio navigation signals degraded on the Earth dayside for about an hour.	140 days per cycle
R2 - moderate (M5 flare)	Limited radio communication blackout and radio navigation degradation on the Earth dayside for tens of minutes.	300 days per cycle
R1 - minor (M1 flare)	Minor radio communication degradation (occasional loss of contact) and degraded radio communication on the Earth dayside for brief intervals.	950 days per cycle

Table of radio blackout effects. The R scale is a scale of the radio blackout effects, whilst the solar flare class is the physical size of the flare on the Sun producing the radio blackout on the Earth. The average frequency is per solar cycle (lasting around 11 years).

stronger flares have happened in the more distant past. Ground level enhancements (GLEs) of solar energetic particles, which only occur during the largest solar flares, alter the chemistry of the atmosphere very slightly. Upon the collision between a solar energetic particle and a neutral atom in the atmosphere, an air molecule can change from one element to another. Take for example, nitrogen, which contains 7 protons and 7 neutrons and accounts for about 78% of the Earth's atmosphere. Upon

An example of an ice core drilled in the Greenland Arctic.

Tree rings.

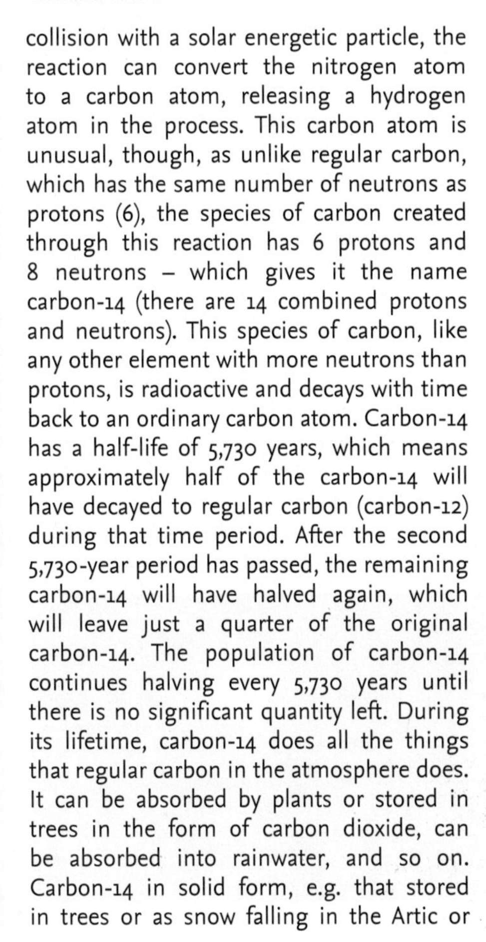

collision with a solar energetic particle, the reaction can convert the nitrogen atom to a carbon atom, releasing a hydrogen atom in the process. This carbon atom is unusual, though, as unlike regular carbon, which has the same number of neutrons as protons (6), the species of carbon created through this reaction has 6 protons and 8 neutrons – which gives it the name carbon-14 (there are 14 combined protons and neutrons). This species of carbon, like any other element with more neutrons than protons, is radioactive and decays with time back to an ordinary carbon atom. Carbon-14 has a half-life of 5,730 years, which means approximately half of the carbon-14 will have decayed to regular carbon (carbon-12) during that time period. After the second 5,730-year period has passed, the remaining carbon-14 will have halved again, which will leave just a quarter of the original carbon-14. The population of carbon-14 continues halving every 5,730 years until there is no significant quantity left. During its lifetime, carbon-14 does all the things that regular carbon in the atmosphere does. It can be absorbed by plants or stored in trees in the form of carbon dioxide, can be absorbed into rainwater, and so on. Carbon-14 in solid form, e.g. that stored in trees or as snow falling in the Artic or Antarctic, will decay to carbon-12 just as it would in the atmosphere. With the average rate of atmospheric carbon-14 production known, the ratio of carbon-14 to carbon-12 in a solid material can be used to date its age. This process, called carbon dating, is used to date wooden products, such as wooden items created by past humans, up to 50,000 years old.

As a tree ages, its trunk will expand only in the summer as the tree grows. This is followed by winter, a period where little tree growth takes place. This annual cycle of growth and less growth within a tree's trunk creates tree rings in the cross section of the tree. Each ring on the tree represents one year of the tree's life. By measuring the evolution of carbon-14 across the tree rings, we can calculate a timeline of how the creation of carbon-14 in the atmosphere changed with time. Because carbon-14 is usually created at a constant rate, any enhancements must be due to an external factor. One such factor is the occurrence of large solar flares. A similar methodology can be used by drilling ice cores in the Artic and Antarctic, where snow fell thousands of years ago but never melted. Just like tree rings, ice cores can tell us about historic atmospheric chemistry, with deeper ice providing insight to further back in time. Ice cores also capture additional elements beyond carbon, such as beryllium and chlorine, which can provide the equivalent proxy over even longer timescales.

By analysing ice cores and tree rings for several decades, scientists have found evidence of historic enhancements in solar energetic particles beyond anything we've observed in the modern age. These enhancements were determined to be due to large solar flares, a handful of which have been discovered in the range of 1,000 to 10,000 years ago. These solar flares would have been at least ten times larger than any flares observed with modern instrumentation and probably even larger than the solar flare triggering the Carrington Event in 1859. When preparing for any natural disaster, however, governments, scientists, and communities need to be realistic about what probability of event it is worth providing the money to prepare for. For example, if preparing for a once in 200-year event costs a particular sum of money, can you always justify spending ten times that to prepare for a once in 2000-year event? There is no correct answer to this question, of course, but it is something that governments and other funders have to consider.

Exosolar Space Weather

Magnetic reconnection is a fundamental process in both solar flares and the formation of the aurora on Earth. In fact, magnetic reconnection is a fundamental process in physics, occurring everywhere where plasma exists. This includes stars, planets, nebulae, and even black holes. By studying processes on the Sun, we are using it as a plasma physics laboratory, which allows us to apply the lessons learnt to stars and other plasma environments all over the universe. Resolved observations (where we can observe the target with more than one pixel) of these environments are rare, but high-resolution images of the Sun can be taken multiple times a minute – the Sun has a lot to teach us. Here are two recent examples of how solar physics has provided insight into other fields of astronomy.

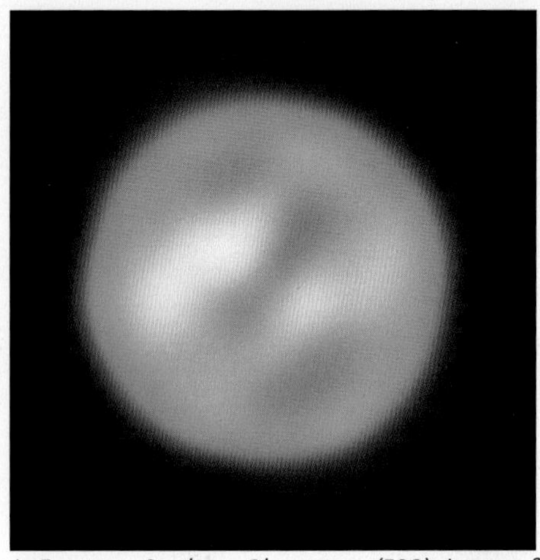

A European Southern Observatory (ESO) image of Antares - the highest resolution image of a star beyond our Sun.

Change in Betelgeuse's brightness between January and December 2019 measured by ESO's Very Large Telescope.

Betelgeuse is a red supergiant and one of the brightest stars in the night sky. Betelgeuse is found in the constellation of Orion, in the upper left corner (or shoulder) of the constellation. Please note, that the 'upper left' of the constellation may not be upper left in the sky for you, depending on where you are in the world. As a red supergiant, Betelgeuse is already approaching the end of its life. Despite being only 10 million years old, the star is expected to go supernova soon, some time between tomorrow and 100,000 years from now. Because of this, Betelgeuse's brightness is closely monitored to search for dramatic brightenings or dimmings that may precede the end of the star's life. Betelgeuse's brightness fluctuates naturally, but in the autumn of 2019 it suddenly decreased to record lows, falling to about a third of its normal brightness. By February 2020, Betelgeuse had fallen in the ranks from the 10th brightest star in the sky to the 25th. The strange behaviour puzzled astronomers, until they realised that the star's true brightness had not decreased at all. Astronomers learned that Betelgeuse had produced a large coronal mass ejection angled so that it blocked out starlight from the star from our perspective. If we'd never observed a CME from the Sun, the puzzle of what happened to Betelgeuse would still be a mystery today.

Research on the Sun's magnetic activity has also provided us with insight into the habitability of exoplanets, planets orbiting around a star other than our own Sun. The first exoplanet was discovered in 1992, and now there are over 5,312 confirmed discoveries as of March 2023 (by the time you read this, that number will have increased). On average, we think that every star in the galaxy has at least one planet orbiting it. For reasons of human curiosity, potentially habitable planets are of particular interest to scientists. A habitable planet is usually defined as such on the basis of its size and distance from its star – two parameters relatively easy to measure. If the exoplanet is at the right distance from its star to offer a temperature suitable for liquid water, it is usually considered 'habitable'. This zone around a star, where temperatures are not too hot (water would boil) or too cold (water would freeze) is called the *goldilocks zone*. The goldilocks zone is not perfect, as it does not consider the atmosphere of a planet. If we were studying our own solar system from a distance, Venus would appear

Artist impression of the Trappist-1 exoplanet system (planet sizes are to scale with one another. The largest planets are comparable in size to Earth).

in the goldilocks zone, but Earth would not. The most famous exoplanet system is TRAPPIST-1, a system of seven planets closely orbiting a red dwarf star. Such a system may seem perfect for life, with multiple planets sitting in the goldilocks zone of a small star that will live for tens of billions of years. However, because of studies of the Sun, we know that solar activity on a dwarf star could create problems. Because red dwarf stars are small, their magnetic field has a strong concentration at the star's surface (far stronger than the Sun's). This will lead to larger sunspots, bigger solar flares, and faster CMEs for a red dwarf – bad news for planets orbiting very closely, as is required to be in the goldilocks zone of such a small, cool star. Although the exoplanets of the TRAPPIST-1 may have habitable temperatures, the high activity on their star means that stellar radiation likely renders the planets completely uninhabitable. As exoplanet science continues to become increasingly popular within the astrophysics community, further applications of solar physics to distant stars will play a big role moving forward.

We aptly refer to solar flares on other stars as stellar flares, the largest of which are observed to come from these red dwarf stars. These *superflares*, as they're sometimes called, are hundreds or even thousands of times more powerful than any solar flare measured directly on our own Sun. Superflares are still far stronger than the larger, more infrequent solar flares detected indirectly through ice core and tree ring analysis.

Open Questions in Solar Physics

Over the last century, we've learnt a lot about the Sun and its influence on us here on Earth. However, our knowledge of the Sun is far from complete, and our local star still holds many mysteries. Around the world, a global community of solar physicists works to increase our understanding of the Sun, making new progress every day. Solar physicists compare cutting-edge observations of the Sun with predictions made by computer simulations or mathematical theory to try and deduce what's truly going on in our local star. The largest breakthroughs in solar physics often come from new tools. This could be a new telescope which observes the Sun in a unique way or at higher spatial and time resolutions than available previously. It could also come from the development of new computer models and simulations which get progressively more and more advanced as computing power increases. Although the list of open questions in solar physics is long, here are the three main areas of mystery (although some scientists may disagree slightly with this list):

1. ***Why is the solar corona so hot?*** The temperature at the Sun's core is a whopping 15 million degrees, powered by the nuclear fusion of hydrogen into helium. The surface of the Sun, the photosphere, has a much lower temperature of 5,500°C. But bizarrely, above the photosphere in the corona, ambient temperatures are as high as 1 million degrees. This temperature discrepancy between the photosphere and corona is known as the *coronal heating problem* and is one of the largest mysteries in solar physics. There are theoretical mechanisms through which the corona can be heated to these high temperatures, but they have not yet been proven observationally. The two leading theories are heating by either nanoflares or *wave-heating*. The nanoflare hypothesis suggests there are tiny solar flares too small for us to observe occurring throughout the solar atmosphere. Although the energy output from each nanoflare is inconsequential, the sheer number of them could produce a total energy large enough to heat the

corona. Alternatively, the wave-heating hypothesis argues that waves travelling along coronal magnetic fields heat the surrounding plasma via vibrations. Both ideas work theoretically (and may each occur in some capacity), but we do not know for certain.

2. ***How are eruptive events created?*** This book has discussed, at length, eruptive events on the Sun (e.g. solar flares and coronal mass ejections), and their impact on Earth via space weather. We are doing a good job at understanding the general and large-scale behaviour of these phenomena, but the exact mechanisms and processes of events triggering their onset is not yet fully understood. Understanding what causes a flare or CME to occur is crucial to eventually forecasting their occurrence, but this too remains a key area of ongoing research.

3. ***What determines the Sun's dynamo?*** The Sun's magnetic field, the cause of all the mischief in the solar atmosphere, is created by the solar dynamo. But how exactly the solar dynamo works or how it is created is not fully understood. The change in the solar dynamo's behaviour with time is also a mystery, and why these variations manifest themselves as both the 11-year solar cycle and further cycles over even longer timescales (e.g. during the Maunder minimum) is also unknown. Like the other open problems in solar physics, we have potential solutions, but no definitive proof just yet.

Golden Age of Heliophysics

It's a good time to be a solar physicist. We have a wide range of instruments available to us, including telescopes on the Earth, telescopes orbiting the Earth, and spacecraft cruising through the solar system. This current era has been nicknamed 'the golden age of heliophysics'. This section introduces three older solar physics missions, which are still used in research and space weather forecasting, and three new projects at the forefront of cutting-edge Sun science.

Solar and Heliospheric Observatory (SOHO): SOHO is a titan of observing the Sun, a collaboration between NASA and the European Space Agency (ESA). Launched in 1995, SOHO was meant to operate for only three years but is still taking measurements of the Sun's atmosphere today. SOHO sits 1.5 million km from Earth, about 1% of the Sun–Earth distance from us. Ambitious

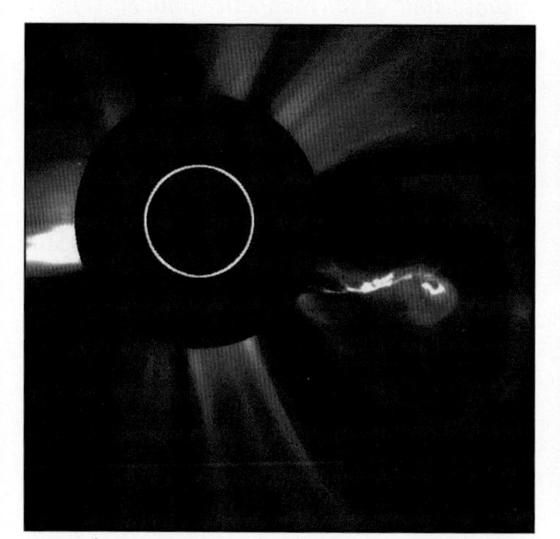

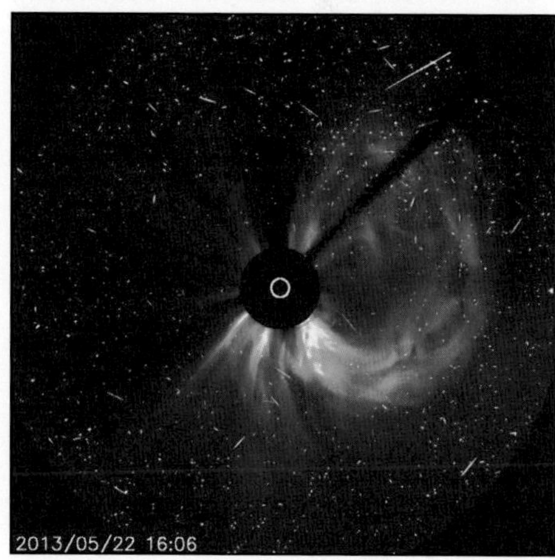

Coronal mass ejections observed by the SOHO LASCO instrument.

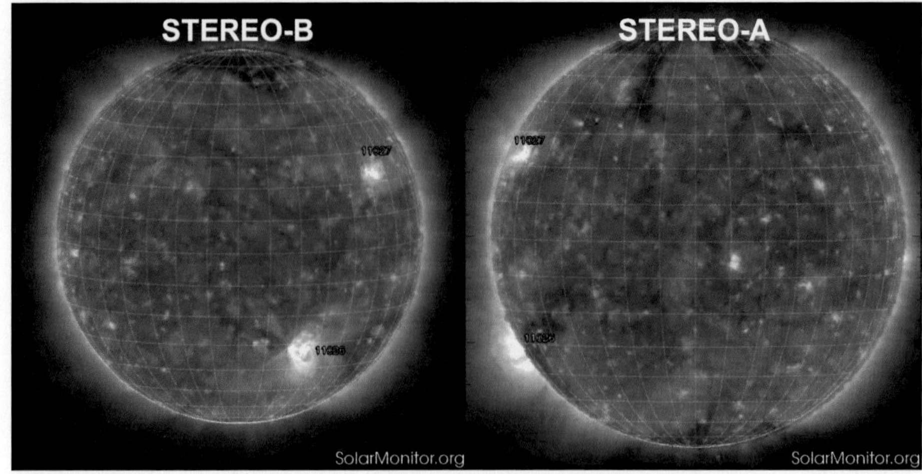

Simultaneous observations of the Sun from multiple viewpoints with STEREO A and B.

missions like SOHO contain multiple instruments with different purposes. The main instrument behind SOHO's continuing success is the *Large Angle and Spectrometric Coronagraph* (LASCO). A coronagraph is a small disc in the centre of a telescope, used to block out the Sun. By blocking out light from the Sun, LASCO is able to observe the much larger solar corona. With two different-sized cameras (LASCO C2 and C3), LASCO is perfectly designed for observing CMEs as they erupt away from the Sun's surface.

Solar Terrestrial Relations Observatory (STEREO): STEREO is a pair of identical spacecraft launched by NASA in 2006. The two spacecraft were put in orbits close to Earth, but with one slightly ahead *(STEREO-A)* and one slightly behind *(STEREO-B)*. Due to the difference in orbits, over the years that followed, the gap between the two spacecraft and Earth increased. Between the two spacecraft, *STEREO* provided a stereoscopic view of the Sun from multiple vantage points. By 2014, the two spacecraft had drifted nearly 180° from the Earth and were about to pass behind the Sun. Unfortunately, NASA lost contact with the *STEREO-B* spacecraft, never to regain full communication again. *STEREO-A* is still in operation today and has drifted so far around the Earth's orbit that it's nearly back at Earth again. The image above shows simultaneous observations from *STEREO-A* and *B*, observing the same active regions from near opposite sides of the Sun.

Solar Dynamics Observatory (SDO): Launched in 2010, the NASA Solar Dynamics Observatory has revolutionised solar physics research and space weather forecasting. SDO has three instruments, including the *Atmospheric Imaging Assembly* (AIA) and *Helioseismic and Magnetic Imager* (HMI). Images from both of these instruments have already featured in this book, including on the front cover. HMI measures the magnetic field of the photosphere, which allows us to study magnetic evolution and *sunquakes* (the Sun equivalent of earthquakes) on the Sun's surface. AIA observes the Sun through ten different broadband filters. The different filters block all incoming light, except for a small wavelength range. The filters are chosen to sample emission lines from different ions on the Sun, each of which forms at a different temperature. Combining these images together, AIA is able to observe plasma at different temperatures on the Sun, which allows

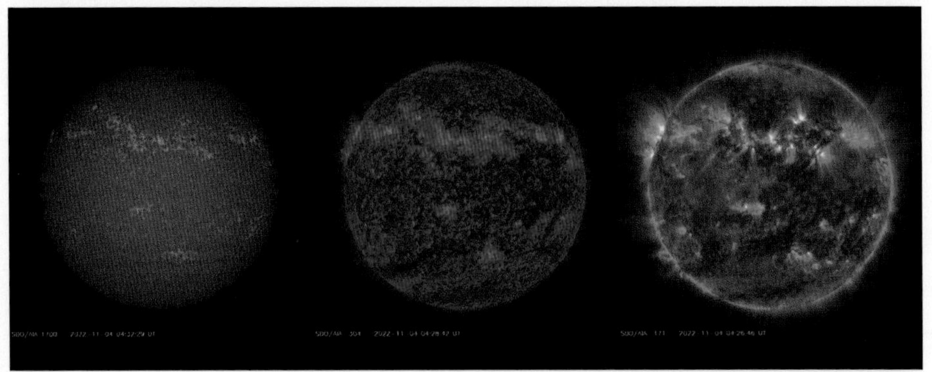

SDO observations of the upper photosphere, upper chromosphere, and corona simultaneously.

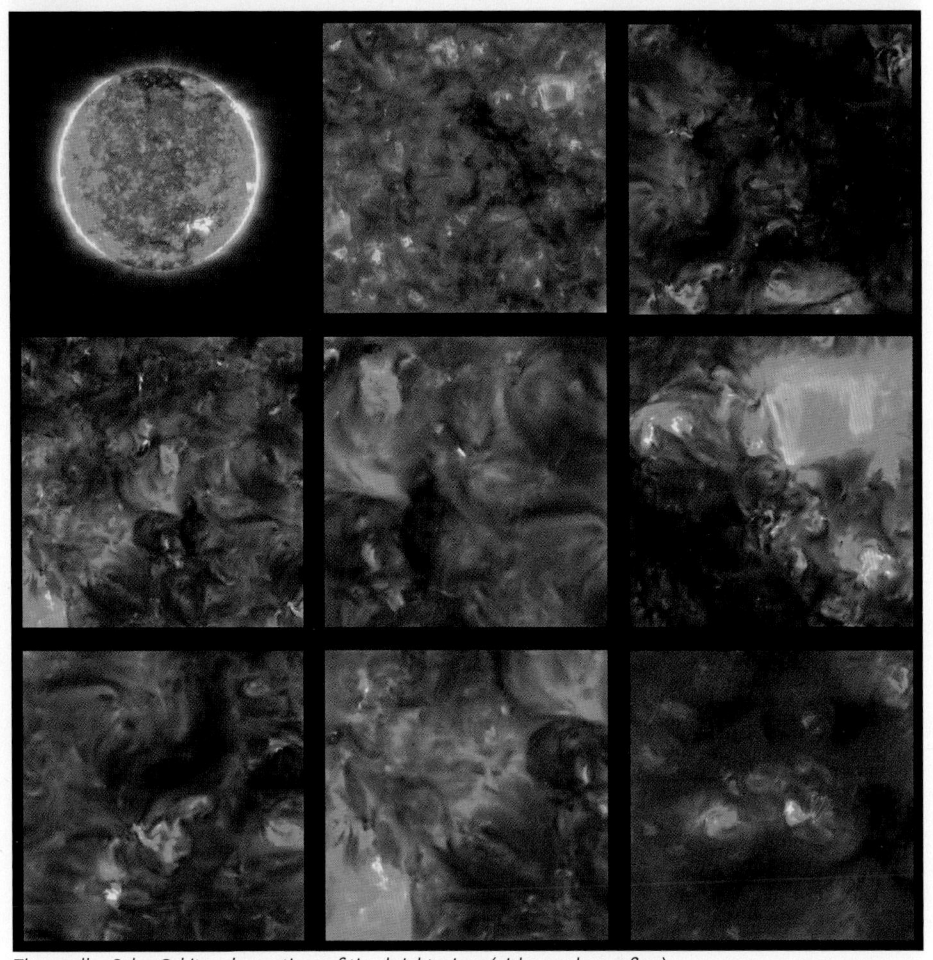

The small – Solar Orbiter observations of tiny brightenings (nicknamed campfires).

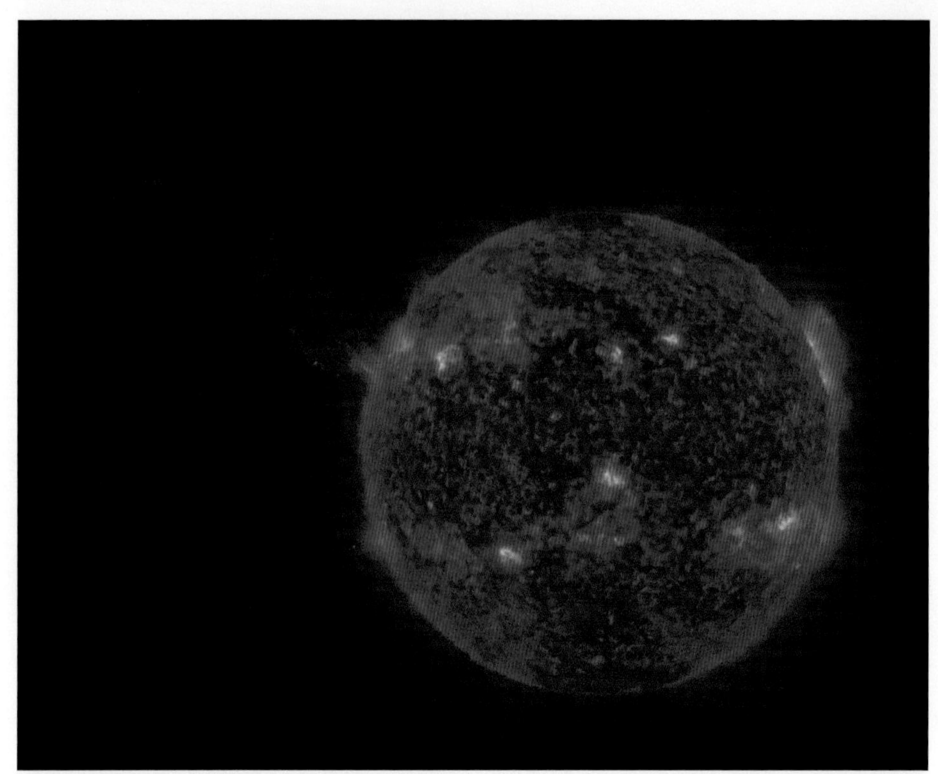

The big – Solar Orbiter observations of a large eruption.

Launch of the Solar Orbiter spacecraft from Cape Canaveral, Florida.

simultaneous images of multiple heights in the Sun's atmosphere. AIA images and how you can access them at home are explored further in the next chapter.

Solar Orbiter: The ESA *Solar Orbiter* spacecraft launched in February 2020, beginning its long journey around the Sun. *Solar Orbiter* contains a number of instruments, some of which observe the Sun, whereas others measure the solar wind environment around the spacecraft. *Solar Orbiter*, as the name suggests, orbits the Sun. Using gravity assists from the planet Venus, *Solar Orbiter* entered an elliptical orbit, which means its distance from the Sun changes significantly. At the closest approach, called the *perihelion*, *Solar Orbiter* observes the Sun from less than 45 million km away, 71% closer to the Sun than the Earth. *Solar Orbiter's* first close perihelion was in October 2022, and it will experience perihelia about twice a year for the next ten years at least. At the closest approach, *Solar Orbiter* is able to observe very high-resolution images of the Sun's photosphere and corona. *Solar Orbiter's* orbit will also migrate out of the plane of the Sun and Earth, which will provide a view of the Sun's poles for the first time.

Parker Solar Probe: The NASA *Parker Solar Probe* launched in 2018, with the goal of measuring the solar wind as close to the Sun as possible. Unlike *Solar Orbiter*, *Parker Solar Probe* does not have cameras to observe the Sun, but measures the magnetic field and plasma parameters around the spacecraft at much closer distances to the Sun than *Solar Orbiter*. At its closest approach to the Sun in 2024 and 2025, *Parker Solar Probe* will be 95.4% closer to the Sun than the Earth, passing less than 5 solar widths above the Sun.

The National Science Foundation's Inouye Solar Telescope near the summit of Haleakalā in Maui, Hawaii.

Inouye Solar Telescope: The National Science Foundation's Inouye Solar Telescope is the only telescope from this list not in space. It is instead on the ground, near the summit of Haleakalā in Maui, Hawaii. In space, telescopes do not have to worry about the Earth's atmosphere. They can observe wavelengths of light blocked by the Sun, not be restricted to observe during the daytime, and not experience any distortion from gases in the air. But, there is a limit to how big a space telescope can be. However, on the ground, telescopes can be built much larger. The Inouye Solar Telescope has a primary mirror 4 metres in diameter and is the largest solar telescope ever built. The previous record holder was the Swedish Solar Telescope located in La Palma in the Canary Islands, with a mirror of 1.5 metres in diameter. Large telescopes are placed high up on mountains and volcanoes where the air is thin, in locations like the Canary Islands, the Atacama Desert in Chile, and Hawaii. The Inouye Solar Telescope began science operations in 2022 and will observe the photosphere and chromosphere of the Sun with unprecedented resolution, able to identify features on the Sun as small as 20 km. The telescope will also be able to measure velocities and magnetic fields on the Sun with very high precision. The image on the page opposite shows the first sunspot observation released by the Inouye Solar Telescope.

Solar Orbiter, Parker Solar Probe, and the Inouye Solar Telescope have not reached the main phase of their missions. Expect to see news stories of discoveries from these projects in the years to come.

This is just a small selection of the current and upcoming facilities available for observing the Sun. The images used in this book are mostly captured in ultraviolet, infrared, or optical light, which provide stunning images of our local star. But observations at these wavelengths provide more than just images – they also give detailed spectra of the Sun to reveal information on plasma temperatures, velocities, magnetic fields, and so on. Understandably, these types of data are not as visually pleasing as the images contained within these pages. Many other operational observatories measure gamma rays, X-rays, microwaves, and radio waves from the Sun, but measurements at these wavelengths do not provide easy to interpret images of the Sun, so are not included here. Other dedicated space missions to study the Sun, including those led by teams in Japan, China, and India, will be launching within the next decade.

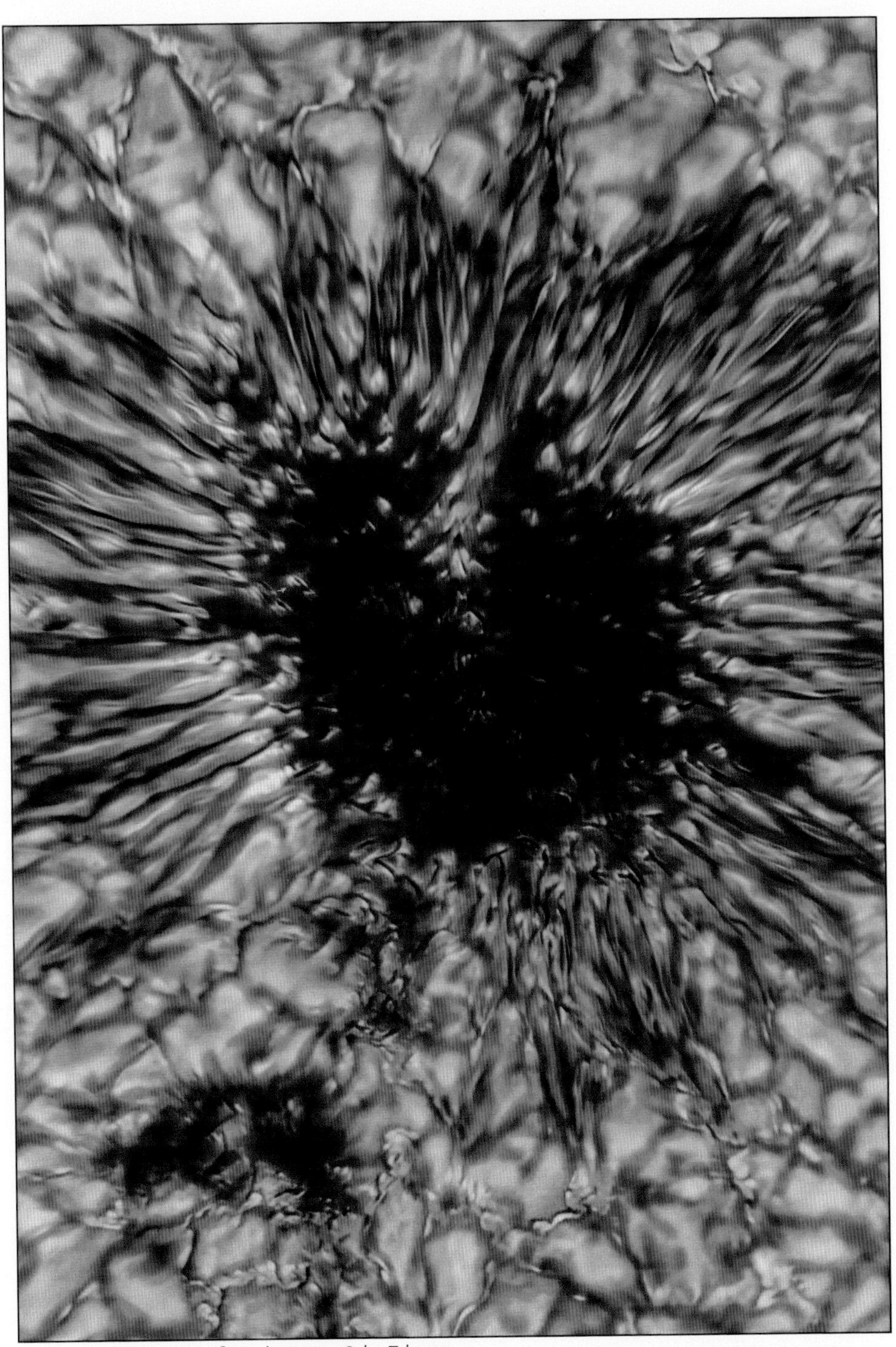

First sunspot observation from the Inouye Solar Telescope.

3: OBSERVING THE SUN

This next chapter is dedicated to how you can observe the Sun. We'll cover how you can observe the Sun safely from home and how to access solar observations from telescopes in space. We'll include some initial information on observing solar eclipses, but Chapter 5 expands on eclipses in more detail.

Observations of the Sun are at the mercy of the 11-year solar cycle. At solar maximum, sunspots, solar flares, and coronal mass ejections are plentiful but can disappear for months at a time during solar minimums. Solar Cycle 24 ended in December 2019, giving way to Solar Cycle 25. The peak of Solar Cycle 24 was smaller compared with those of previous solar cycles, with lower sunspot numbers and fewer flares. At the time of writing, we are in the rise phase of Solar Cycle 25, but it is not known how large the cycle will be or when its maximum will occur. Some scientists predict that Solar Cycle 25 will have a peak smaller than that of Solar Cycle 24, but others predict a larger peak closer to those of previous cycles 22 and 23. Depending on when you read this book, we may have reached the peak of Solar Cycle 25. It is important to consider the phase of the solar cycle when observing the Sun. If you buy equipment to observe sunspots during a solar minimum, you may be disappointed. Between 2023 and 2028, sunspots will likely be plentiful over the Sun's surface. However, if you are reading this later than 2028, there's a chance that solar activity is already decreasing towards the next solar minimum, providing fewer features of interest to see on the Sun. Towards the end of the 2020s or early 2030s, Solar Cycle 25 will end and give way to Solar Cycle 26. The dates of this are approximate, depending on the timing of the peak of Solar Cycle 25. Before heading outside to observe the Sun, your best bet is to look online first to see whether the Sun is doing anything worth observing.

Safety Tip

Remember, with one specific exception, you should never look at the Sun directly. The effects of gazing directly at the Sun with the unaided eye may include a changed colour perception, blind spots, distorted vision, and so on. The effects are usually temporary, and your eyesight will revert to normal after a few minutes of being away from the Sun, in a darker location. If symptoms persist for more than a day, seek medical attention from your doctor. The only time it is safe to look at the Sun is when its entire surface is blocked by the Moon during a total solar eclipse. The few minutes of eclipse totality (not before or after during the partial phase) are the only time it is safe to look at the Sun with your unaided human eyes. But please note, as we will explore in more detail during this chapter, eclipses are rare, and only safe to look at directly for the few minutes of totality within a small geographical region. If you are unsure whether such conditions are present or applicable, then they are likely not. In which case, I reiterate once more – do not look directly into the Sun.

Seeing through an Atmosphere

The Earth has an atmosphere. We rely on the atmosphere to live, but not only because it keeps us warm and provides us with air to breathe. The atmosphere also does a fantastic job of blocking out wavelengths of light harmful to life on Earth. High-energy wavelengths of light – ultraviolet, X-rays, and gamma rays – are forms of ionising radiation which can damage DNA in cells and lead to the growth of cancer and other mutations. Luckily for us, most harmful radiation does not make its way to the Earth's surface. Instead, it's absorbed within the stratosphere and above (at heights over 20 km above us). The *ozone layer* is one specific layer of the stratosphere and contains a high concentration of a variety of oxygen molecule called ozone, which is especially efficient at absorbing harmful ultraviolet light. All in all, the Sun's atmosphere blocks gamma rays, X-rays, and most ultraviolet wavelengths from reaching the Earth's surface. Less harmful longer wavelengths in the infrared and microwave regions of the spectrum are also blocked. Only optical light, near ultraviolet, near infrared, and long-wavelength microwaves and radio waves are able to pass through the atmosphere completely. It is not a coincidence that optical light, the light our eyes are sensitive to, can pass through the atmosphere, as our eyes evolved specifically to see the light available to them.

What was good for the evolution of life is less ideal for astronomy. The majority of images of the Sun shown in this book show the beautiful solar corona seen in ultraviolet light – the same wavelengths of light that cannot pass through the atmosphere above us. In order to observe X-ray and extreme ultraviolet measurements of the Sun, which are critical for understanding high-energy events in the Sun's atmosphere, we must put telescopes in space, above our own atmosphere. This problem is not exclusive to solar physics either, as ultraviolet and X-ray measurements are important for many fields of observational astronomy. Down on the surface, we are limited to observing the Sun at wavelengths not blocked by the Earth's atmosphere and observed by telescopes sensitive to optical light, infrared, and microwave/radio. Your human eyes are sensitive only to the optical range of light (red through blue), and so the only regions of the Sun you can ever hope to observe from home are limited to those regions producing light in this wavelength range. Therefore, although you can observe the photosphere and chromosphere, you will not see the solar corona (again, unless during a total solar eclipse). To enjoy the beauty of the Sun's corona, you will have to utilise open-source observations from space telescopes. These images are readily available online, with the steps for accessing them also outlined in this chapter.

Even for observations in white light, the atmosphere is still a nuisance, constantly in motion regardless of the presence of clouds. The air above us churns and convects, continuously wobbling between us and the Sun or stars we wish to observe. The term *seeing* is a measurement of the quality of sunlight or starlight reaching us on the ground after passing through the atmosphere. Essentially, the more air there is between us and space, or the more turbulent that air is, the worse the seeing. Professional observatories are constructed in high-up, dry locations, where the air is thinnest, to provide the best seeing possible. Bear this in mind when observing from home. If it is particularly windy, humid, or the Sun is lower in the sky during early morning or late afternoon, there is more air between you and the Sun, providing worse seeing than on still, windless days with the Sun directly overhead. If you are using a solar telescope or other viewing device to observe the Sun from home, the level of seeing will significantly impact the clarity of your observations and the size of the features you wish to target.

An enhanced/extended range of available wavelengths and the absence of atmosphere are the two primary advantages of putting telescopes in space. However, ground-based telescopes have advantages too. They are far cheaper than equivalent-sized space-based telescopes, can be built far larger, and can be easily maintained, modified, and upgraded throughout their lifetime. But fear not – you do not need a state-of-the-art ground-based observatory to spot features of interest on the Sun.

Solar eclipse glasses.

Eclipse Glasses

Eclipse glasses are the simplest way to observe the Sun safely, but there are limits to what you can see through them. Eclipse glasses are usually made of cardboard in the shape of regular glasses, with filters placed where typical glasses would have lenses. Other eclipse glasses designs do not use the 'glasses' shape but are handheld rectangles with the filter in the centre. Eclipse glasses use filters of either a black polymer or silvery mylar, each blocking out 99.9999% of sunlight. You should never use regular sunglasses to look at the Sun, as these only block out around 20% of sunlight in comparison. Because they transmit only a small fraction of light, eclipse glasses give the Sun a colour, depending on their filter. The Sun will appear orange or blue-white through black polymer or silver mylar respectively. As the name suggests, eclipse glasses are perfect for looking at eclipses. During a partial solar eclipse, when the Moon blocks a fraction of the Sun's surface, eclipse glasses provide a clear view of the cosmic show. If the Sun is totally blocked by the Moon, in a moment that only lasts a minute or so over a small geograpical region, eclipse glasses are not needed. Eclipse glasses can also be used to observe large sunspots. Only the largest sunspots can be seen through eclipse glasses, as small sunspots will require some form of

Eclipse filters being used during the 1914 solar eclipse.

magnification to see. Because they are cheap and easy to use, eclipse glasses are a great introductory resource for observing the Sun. When buying eclipse glasses, make sure you use a verified vendor. Eclipse glasses and solar filters have a safety requirement to meet the ISO 12312-2 international standard, which some companies online do not do. If you are unsure, you can visit the Reputable Vendors webpage of the American Astronomical Society (included in the resources page at the end of this book).

Solar Projectors

If you own a regular (night sky) telescope, you can turn it into a solar projector.

Solar projection method used to watch a partial solar eclipse.

Provided you do not look down the telescope eyepiece yourself, this method is perfectly safe. If your telescope is smaller than a few inches in aperture, this will not damage your telescope (the telescope can handle the bright light, even if your eye can't). If your telescope is larger or a Schmidt-Cassegrain, using a filter to block some light will prevent excess heat from damaging the adhesive between telescope components. Be sure to cover the finderscope whilst using your telescope in this way.

To turn your telescope into a solar projector, use a low-power eyepiece (one with no cemented lens) and place a white card 20–60 cm away from it. You'll see the projection of the Sun on the card, which you can move until the Sun is in focus. The exact card–eyepiece distance required will depend on the focal length of your telescope. Shading your projection from regular daylight will increase the clarity of the image. If you're feeling creative, you can create a makeshift mount for your projection surface to allow a longer appreciation of the projected image of the Sun.

There are purpose-built solar projectors which use this same mechanism but with a purpose-built lens. One brand of these is called Sun Spotters. These can be relatively expensive, however, compared with reutilising a night sky telescope if you have access to one.

The projection of the Sun will be in white light, providing a similar (but magnified) view to that with the eclipse glasses. You'll be looking at the Sun's photosphere, with sunspot regions clearly visible. This method of solar projection through a standard telescope is the same method used by Galileo, Carrington, and other historical solar observers. The advantage of the projection method over a purpose-built solar telescope (which would allow you to look through the eyepiece directly) is that larger groups can enjoy the solar projection simultaneously, instead of having a single viewer at the eyepiece at a time.

Solar Telescopes

To observe the Sun best from your back garden, you'll want to use a solar telescope. You can either buy a purpose-built solar telescope or a filter for your night sky telescope. If you're looking to do the latter, make sure you buy the filter from a reputable source. The American Astronomical Society website (included in the resources page at the end of this book) provides a list of verified vendors. Solar filters for your night-time telescope will be white-light filters, which provide a view of the solar photosphere. For purpose-built solar telescopes, there are two main types.

White-light telescopes: White-light solar telescopes work the same as white-light filters for a regular telescope. They reduce the intensity of all light entering the telescope, but do not discriminate by wavelength. As a result, you are seeing white-light emission from the Sun, which originates from the photosphere. These

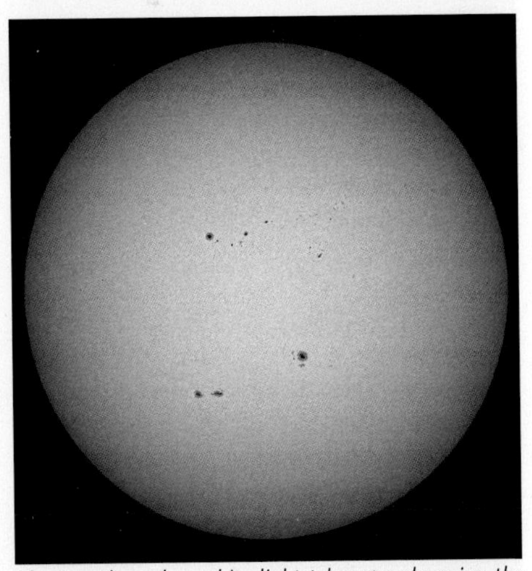

The Sun through a white-light telescope observing the photosphere.

allow you to see sunspots and some surface structure. They are the cheapest kind of solar telescope.

H-alpha telescopes: The second kind of solar telescopes are ones which block all light, apart from light in a very narrow wavelength range. This wavelength range is selected to include a strong emission line on the Sun, observing light emitted during one specific electron transition. By observing light produced by a specific transition, we see the layer in the Sun creating these transitions, usually at different heights in the Sun beyond the photosphere. The most useful wavelength to observe is Hydrogen-alpha, or H-alpha. Because H-alpha emits at a wavelength of 656.3 nm, we see it as red light through the telescope. H-alpha is produced in the solar chromosphere, the layer of the Sun between the photosphere and chromosphere. H-alpha is a broad emission line, and the view of the Sun through an H-alpha telescope will depend on which part of the emission line the H-alpha filter lets through. Most H-alpha telescopes

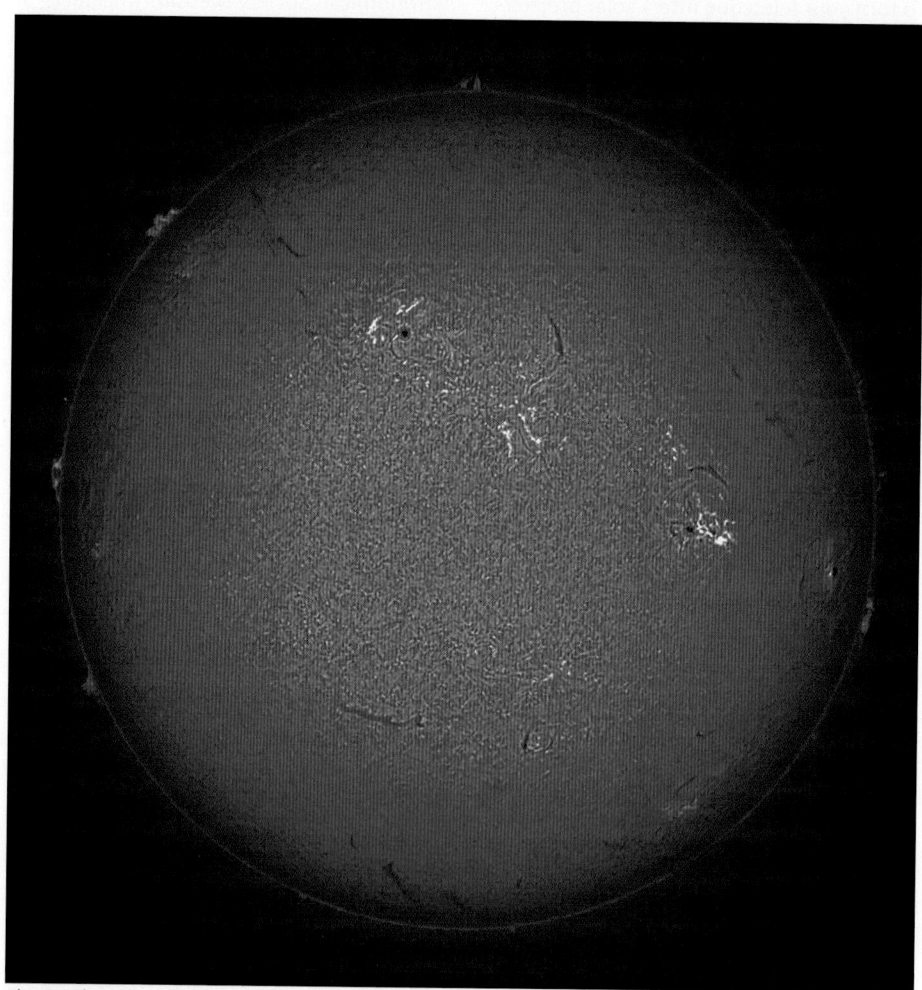

The Sun through a H-alpha telescope observing the chromosphere.

have a 'tuning' function, to slightly modify the wavelength of H-alpha observed by the telescope. Tuning the filter allows us to tune between stationary components of the chromosphere and plasma that has been blue- or red-shifted (usually because it is on the edge of the Sun). Many interesting features exist in chromospheric plasma. These include sunspots, *filaments,* and *prominences* (the next chapter describes these features). If you're lucky, you could even observe a solar flare in H-alpha in the form of flare ribbons and flare loops. Only the very largest flares are visible in white-light telescopes, but far easier to detect in H-alpha observations. Other filters centre on other spectral lines in the visible light range, including Ca K, which is a calcium line also produced in the chromosphere. This line is weaker and harder to see than H-alpha, although it can come out well in photographs.

Pinhole Camera

This next method of observing the Sun is the easiest of them all, but only works during a partial solar eclipse. If you find yourself without eclipse glasses or a solar projector during this event, fear not. Sunlight passing through a small gap will project the shape of the light source, in this case the Sun. Simply make a small hole in a sheet of paper and let the sunlight shine through it onto a flat surface. The shape projected will show you the shape of the eclipse. It's really as simple as that. You don't need to make one either, you can also use a colander or even the gaps between leaves in a tree.

Accessing Space Data from Home

Since the launch of the NASA Solar Dynamics Observatory in 2010, we've received a near-continuous stream of 4k

Leaves on a tree acting as a pinhole camera projecting the partial eclipse onto the ground.

AIA filter	Temperature sensitivity	Region of atmosphere
4500 Å	Photosphere	5,000 °C
1700 Å	Temperature minimum in the photosphere	5,000 °C
304 Å	Chromosphere, transition region	50,000 °C
1600 Å	Transition region, upper photosphere	0.1 million °C
171 Å	Quiet corona, upper transition region	0.6 million °C
193 Å	Corona and hot flare plasma	1.6 and 20 million °C
211 Å	Active-region corona	2 million °C
335 Å	Active-region corona	2.5 million °C
94 Å	Flaring corona	6.3 million °C
131 Å	Transition region, flaring corona	0.4 and 10 million °C

resolution images of the Sun, captured multiple times a minute at ten different wavelengths. Each wavelength observed by the SDO AIA instrument is sensitive to a different temperature and provides images of different layers of the Sun. The table above summarises these ten wavelengths and the temperature/regions of the Sun they observe. The filters used by AIA are given in units of angstroms (Å) equal to 0.1 nanometre. Some filters include more than one emission line, so have sensitivity from more than one temperature.

These images are used by solar physicist researchers and space weather forecasters. However, SDO observations are also easily accessible by the public. This is because SDO, like most NASA missions, is funded by the US taxpayer, with a policy that all images and data produced by publicly funded missions are available to everyone. Other space agencies, such as the European Space Agency (ESA) or the Japanese Space Agency (JAXA), have similar data policies too. Most data from space missions is not visual, but images of the Sun are an exception. The page opposite shows images from all nine of the AIA passbands. You'll notice they seem very colourful. This is not the true colour of the Sun, but because most of the images are in ultraviolet light, they have no true colour at these wavelengths we can see. Instead, NASA utilises a different colour for each filter of the telescope.

There are multiple ways of accessing space observations of the Sun. For SDO data specifically, the data page of the SDO website (included in the resources page at the end of this book) is very useful. The page shows the most up-to-date images of the Sun but also allows you to search by date to create custom images or films. For viewing the most current solar data, solarmonitor.org is also a very useful website, displaying images from other telescopes too. These pages limit the downloadable resolution to 1072p, less than the full 4k resolution of the images. Truthfully, unless you are doing science with the images, the full 4k resolution is not necessary.

The best tool for creating solar images and videos is a piece of ESA-developed software called JHelioViewer. The programme does have an online equivalent, HelioViewer, without the full functionality. The screenshot on the next page demonstrates the

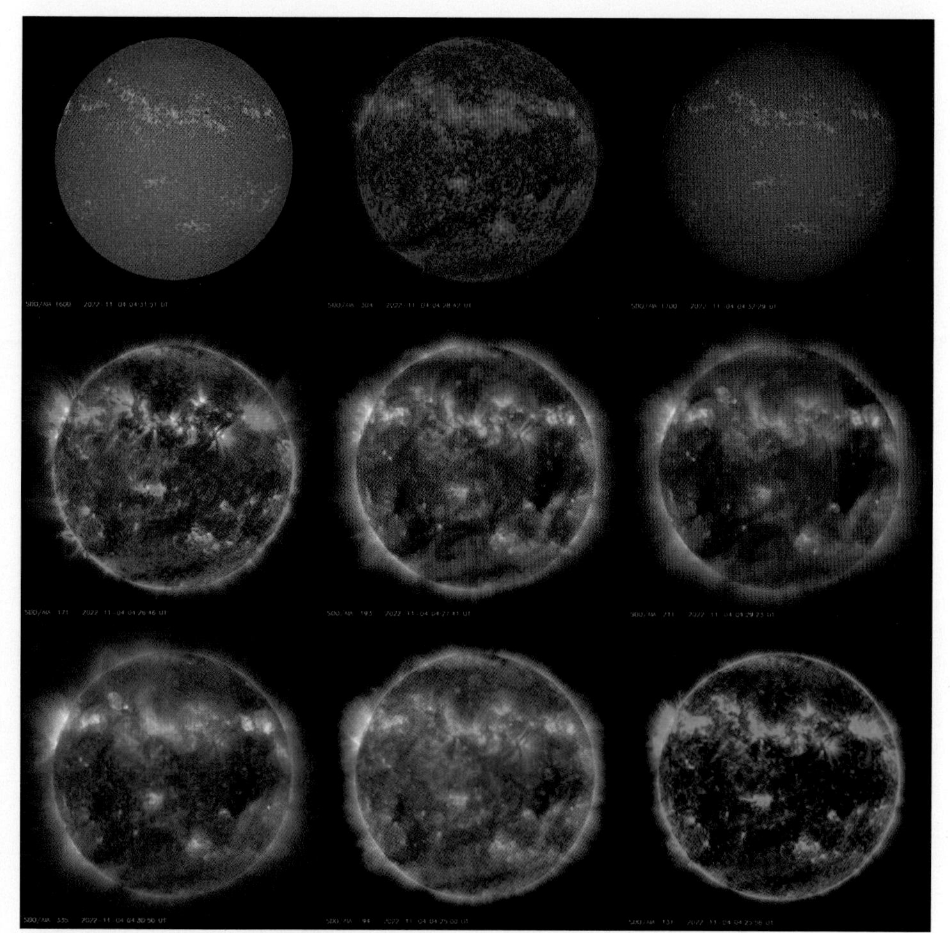

The Sun observed through nine SDO/AIA passbands.

JHelioViewer programme. After selecting your date range, you can select observations from almost every instrument on board every solar spacecraft. You can align and blend these observations and edit the image properties such as colour, brightness, and contrast. JHelioViewer creates high quality videos and includes everything you'll need to explore space observations of the Sun.

Using JHelioViewer and other websites, you can take a look at what the Sun was doing on your birthday, anniversary, and so on. Did the Sun do anything on this special day or was it relatively uninteresting? To demonstrate this, the image on the front of this book is the Sun on the day of my wedding, observed at a wavelength of 171 Å. You'll notice a bright region in the bottom left of the image. This was an X-class solar flare. SDO data is only available from 2010, but lower-resolution images from other spacecraft (like the SOHO EIT instrument) are available from as far back as 1996.

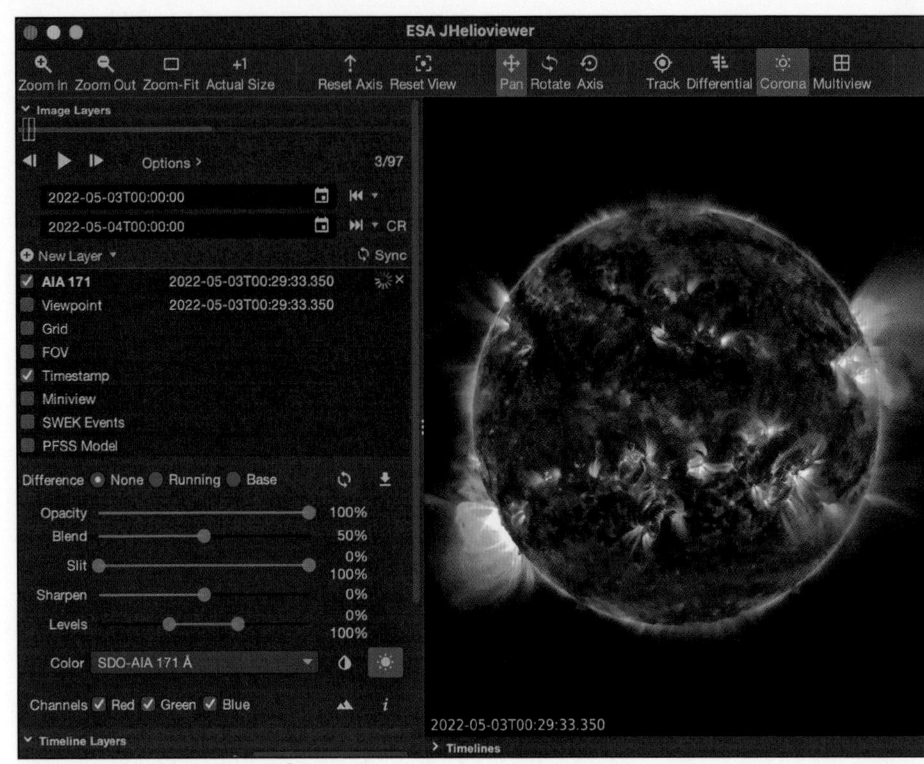

Screenshot of the JHelioViewer software.

4: WHAT TO LOOK FOR ON THE SUN

This chapter summarises a lot of the content we've learnt so far, providing a brief guide to the different observable features on the Sun.

Photosphere

What is it? The photosphere is the layer of the Sun we receive sunlight from. It is the 'surface' of the Sun, at the very base of its atmosphere.

What can I find there? The photosphere is the least dynamic layer of the solar atmosphere, but it is the easiest to observe from home. Features to see there include sunspots, *plage*, and limb darkening (described later in this chapter).

How to view it from home? White-light telescopes, solar projectors, eclipse glasses.

Chromosphere

What is it? The chromosphere is the layer above the photosphere where temperatures rise and densities fall rapidly over relatively short heights.

What can I find there? Sunspots, flare ribbons, flare loops, filaments, and prominences. Although filaments and prominences are not in the chromosphere, they contain chromospheric plasma and so are observed the same way.

How to view it from home? H-alpha telescope.

Corona

What is it? The corona is the main region of the solar atmosphere and is dominated by plasma flowing along magnetic fields.

What can I find there? Active region loops, solar flares, coronal mass ejections, coronal holes, filaments, and prominences.

How to view it? Unless you have very specialised coronagraph equipment, you cannot usually view the corona from home. The only exception to this is during the totality phase of a total solar eclipse. You can find observations of the corona online in extreme ultraviolet (EUV) or coronagraph data.

Sunspots

What are they? Dark regions of the photosphere caused by concentrated magnetic fields suppressing plasma flows. Sunspots have a dark centre, the *umbra*, and a lighter edge, the *penumbra*.

How to spot them from home: Sunspots are the easiest thing to observe on the Sun. Visible with solar telescopes (white-light or H-alpha), solar projectors, and even eclipse glasses for larger sunspots.

How to find them in online data: Most visible in observations of the photosphere or chromosphere, for example, AIA 4500 Å, 1600 Å, or 1700 Å. HMI magnetograms can show the magnetic fields on sunspots.

Active Regions

What are they? Hot plasma flowing along strong magnetic field lines and found above sunspots or plage patches (brighter patches often surrounding sunspots). Active regions are the origin of high energy solar activity, like solar flares or coronal mass ejections.

How to spot them from home: Active regions exist in the solar corona, so they are not observable by standard home-based observing techniques. As is mentioned in the corona section above, they can be observed during a solar eclipse or with special coronagraph equipment.

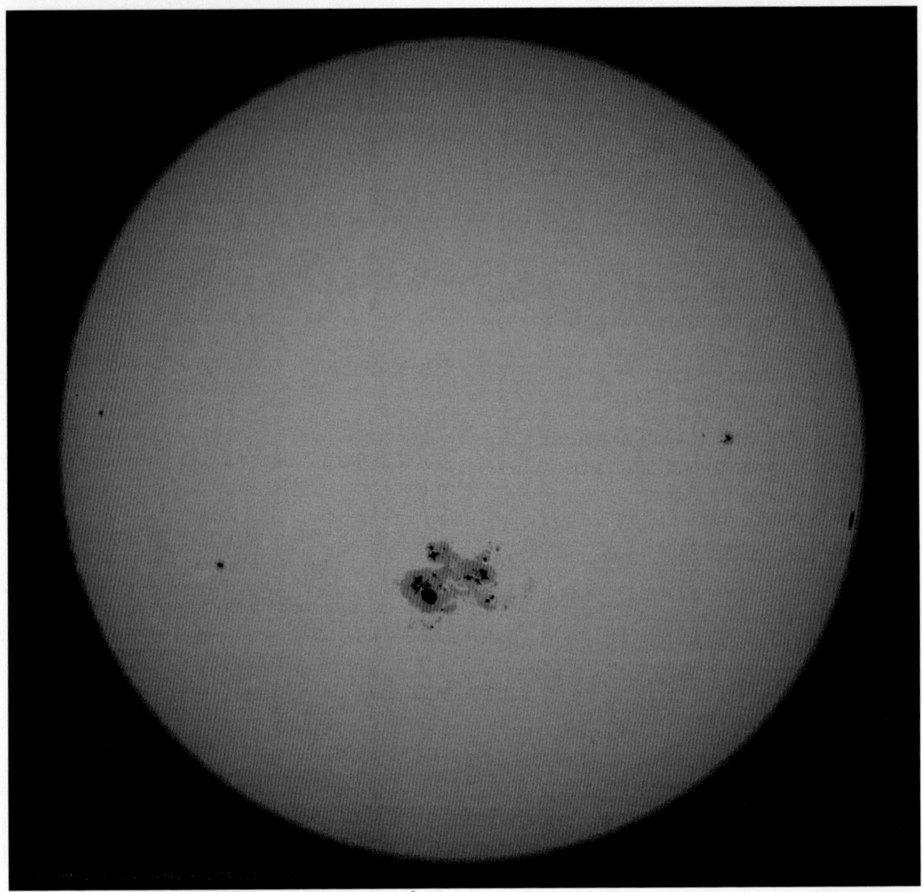

A large sunspot patch observed by SDO/AIA 4500 Å.

How to find them in online data: Visible as bright features in most extreme UV images from SDO/AIA, most noticeable in 193 Å, 131 Å, 171 Å, 211 Å, 335 Å, and 94 Å channels.

Filaments and Prominences

What are they? Filaments are channels of dense chromospheric plasma suspended in the corona by strong magnetic fields. Filaments can sit over active regions or the quiet Sun and can erupt as CMEs. Filaments and prominences are the same thing, but the naming convention is that filaments are visible against the Sun (what we call the disc), whilst prominences are visible over the edge of the Sun (what we call the limb) against the black background of space.

How to spot them from home: H-alpha observations can see filaments and prominences. Due to the high contrast with a black background, prominences are easier to spot, being visible as fuzzy features on the edge of the Sun. Sometimes the legs of prominences are visible, but other times they are not (which makes the filament body appear to 'float'). Filaments are viewed against the Sun and appear as long tendril-like features with a different brightness from that of the background Sun.

Active region observations from SDO/AIA 171 Å.

How to find them in online data: Some filaments can be seen as dark structures in extreme UV images from SDO/AIA and are most visible in the 304 Å channel. However, the best way to spot filaments is in H-alpha measurements from a telescope network called the Global Oscillation Network Group (GONG). On a website called solarmonitor.org, recent H-alpha measurements from GONG are available.

Solar Flares

What are they? Solar flares are the explosive release of energy in the solar atmosphere in the form of plasma heating, particle acceleration, and light. Flare loops are bright loops forming in the corona, with their footprints in the chromosphere marked by flare ribbons.

How to spot them from home: Observing a solar flare from home requires luck. White-light flares are very rare, so you'll need an H-alpha telescope. If you are lucky enough to observe a solar flare, you'll see the bright ribbons in the chromosphere. If you notice online that a solar flare has begun, you may have time to set up your H-alpha telescope and catch the later stages of the event.

How to find them in online data: Solar flares are most visible in SDO/AIA observations at 131 Å. You can look up soft X-ray data (from NOAA) to find the timing of a flare and look for the corresponding observations in AIA images. Catalogues are also available with lists of solar flares.

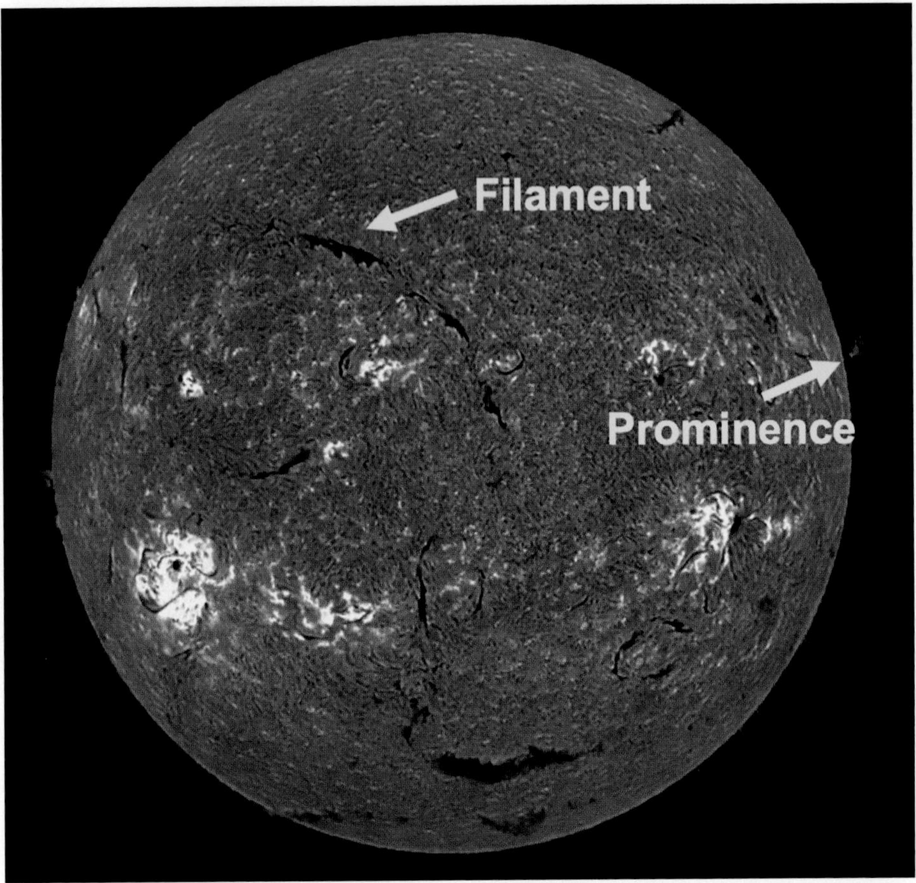

H-alpha observations of prominences and filaments from the Big Bear Solar Observatory.

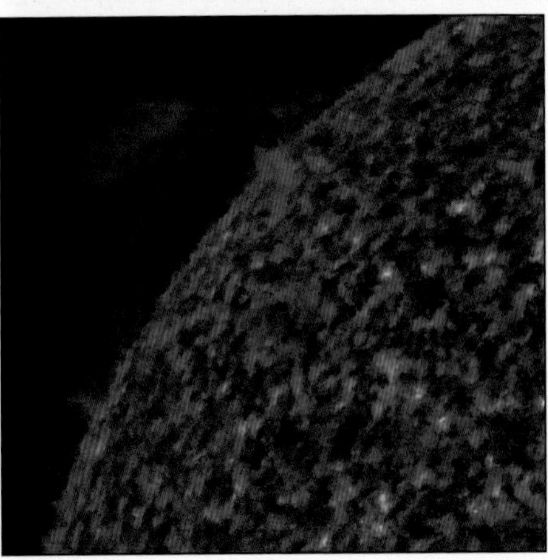

Prominences.

Eruption of a filament.

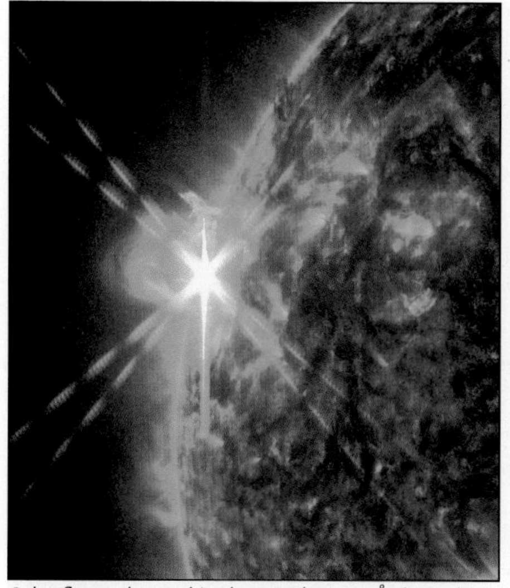

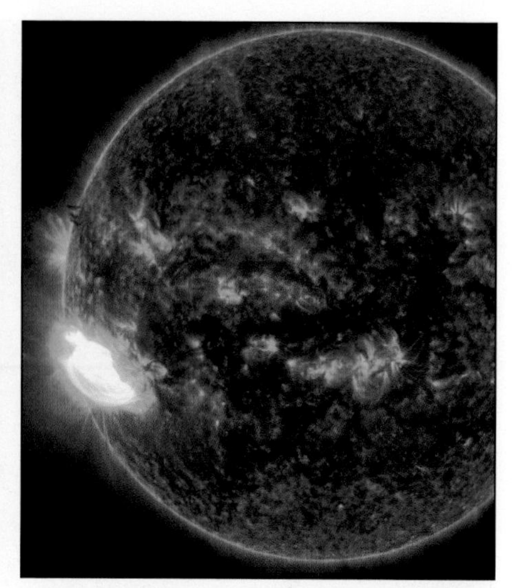

Solar flares observed in the SDO/AIA 131 Å channel.

Coronal Mass Ejections (CMEs)

What are they? CMEs are eruptions of plasma from the solar atmosphere into the solar system.

How to spot them from home: Like other observables in the corona, you cannot observe CMEs from home without a specialist coronagraph or, if you get incredibly lucky, during a total solar eclipse.

How to find them in online data: CMEs are most visible in data from the LASCO instrument on board SOHO. LASCO has two cameras, C2 and C3, which use a coronagraph to block out the Sun's surface

Coronal mass ejections viewed by SOHO/LASCO C2.

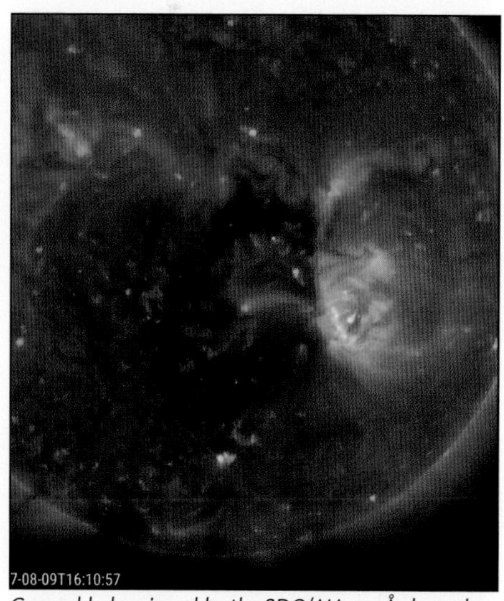

Coronal holes viewed by the SDO/AIA 211 Å channel.

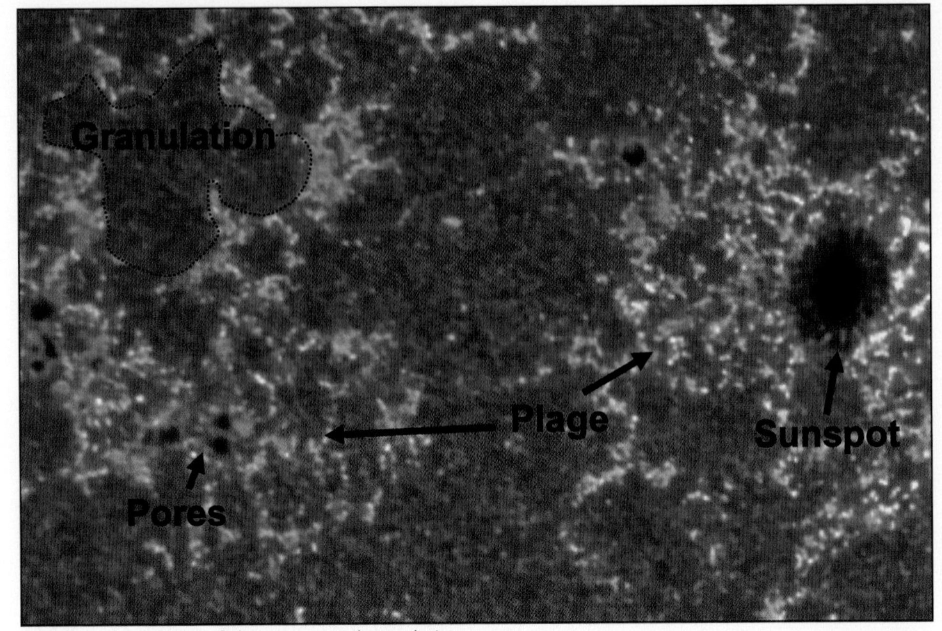

SDO/AIA observations of plage, pores, and granulation.

and view the corona. The two cameras have different fields of view, observing different distances away from the Sun. As the CMEs erupt through the corona, LASCO measures white light scattering off the CME plasma. Ground-based coronagraphs, such as the K-Cor telescope at the Mauna Loa Solar Observatory, can also observe CMEs. Coronagraphs can also observe the regular solar wind emanating away from the Sun.

Coronal Holes

What are they? Coronal holes are regions of open magnetic fields in the solar atmosphere creating a source of the fast solar wind.

How to spot them from home: Coronal holes exist in the corona and are only visible when viewed straight on. Unfortunately, this means that even with a coronagraph or total solar eclipse, you'll never be able to observe these from the ground.

How to find them in online data: Coronal holes are easy to spot in SDO/AIA measurements in the 193 Å and 211 Å channels, being visible as dark regions taking up a large part of the Sun.

Other Small Features, e.g. Plage, Pores, and Granulation

What are they? There are other smaller miscellaneous features visible in the Sun's photosphere. These include plage – slightly lighter patches in the upper photosphere and chromosphere, that don't have a magnetic field strong enough to create sunspots. Lower down in the photosphere, these regions are called faculae. *Pores* are small isolated dark spots, similar to sunspots, but with fewer structures and without a penumbra. *Granulation* is the top of convection cells that extend beneath the Sun's surface.

How to spot them from home: Plage, pores, and granulation are all features in the photosphere, so can be viewed by a white-

Limb darkening observed by SDO/AIA 4500 Å.

light solar telescope. They are small and difficult to spot, especially granulation, so are only visible with larger telescopes.

How to find them in online data: Easily visible in photospheric observations, such as those of SDO/AIA 4500 Å, 1600 Å, 1700 Å, and HMI magnetograms. Surface granulation is always visible, especially in the last three channels.

Limb Darkening

What is it? Limb darkening is the darkened appearance of the Sun around its edges (the limb). When we look at the limb of the Sun, we are not looking as deep into the Sun as when we look head on at the disc. Because of this, we are seeing slightly cooler plasma, which causes the limb to appear darker.

How to spot them from home: Limb darkening is most visible in the photosphere, so a good eye will notice it through a white-light telescope.

How to find them in online data: Limb darkening is easily seen in SDO/AIA 4500 Å, 1600 Å and 1700 Å measurements of the photosphere.

5: SPECIAL SOLAR EVENTS

Transits

For a solar physicist, one of the most interesting things a planet can do is pass in front of the Sun. There are eight planets in the solar system, but only two of them, Venus and Mercury, are closer to the Sun than the Earth. Because their orbits are inside ours, we sometimes see these planets pass in front or behind the Sun from our perspective. As they pass near the Sun, Venus and Mars reflect sunlight and are visible in coronagraph observations from *SOHO/LASCO*. What is far more impressive, however, is when the planets pass in front of the Sun. This is what we call a planetary transit.

The Earth, Venus and Mercury orbit the Sun every 365.25, 225, and 88 days respectively. However, because the orbits of the planets lie close to but not exactly within the same plane, alignments between the Sun, Earth, and Venus or Mercury are rare. Most of the time Venus and Mercury pass above or below the Sun from our perspective – transits are rare. Mercury transits take place 13–14 times a century, coming in clusters of two or three over 3–4 years, followed by a larger gap of 10 years or more before the next cluster of transits. Previous transits of Mercury took place in 1999, 2003, 2006, 2016, and 2019. Unfortunately, the next one will not be until 13th November 2032. Transits last for several hours and can be seen by about a third of the world each time. Mercury appears very small as it passes over the Sun and is visible as a small black dot. You can observe the transit with any solar telescope or solar projector, but the planet is too small to see without magnification, so eclipse glasses will not do the job.

The transit of Venus is a different beast. Venus is far larger than Mercury and much closer to us, so its apparent size in the sky is much larger. During Venus transits, Venus is large enough to see without magnification, so can be seen with eclipse glasses alone. Solar telescopes and solar projectors will still provide a better view, however. The view is certainly impressive, as shown in the image on the page opposite. Unfortunately, providing advice on how to view a Venus

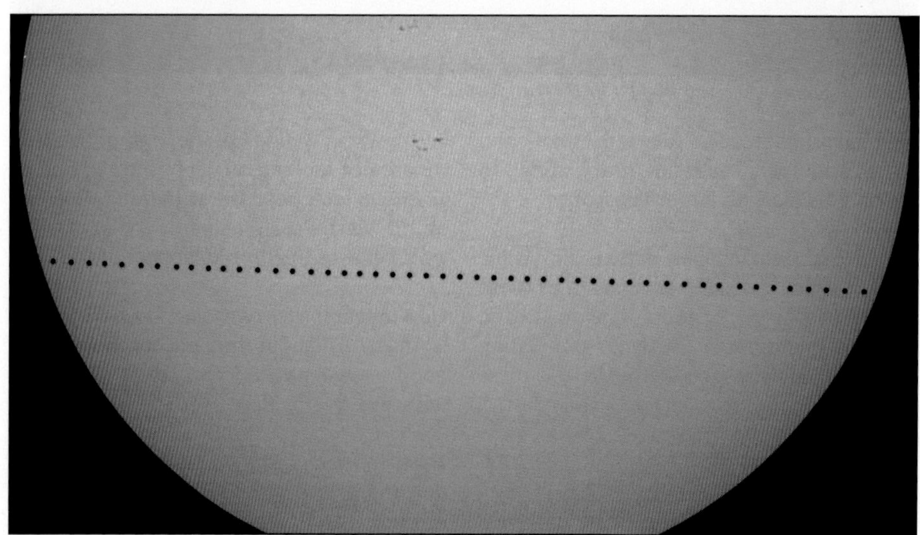

Transit of Mercury.

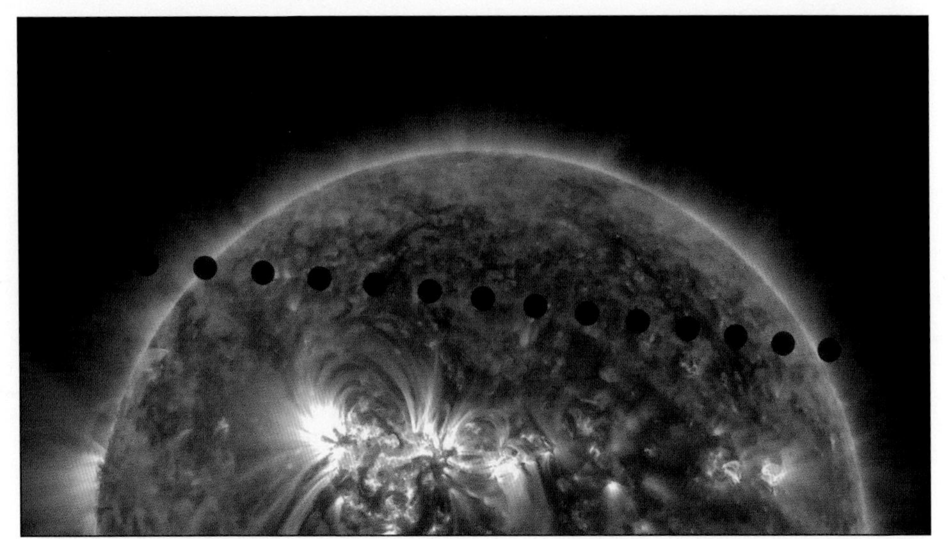

Transit of Venus.

transit seems slightly obsolete. Although the previous two Venus transits were in 2004 and 2012, the next one is not until the 11th December 2117. It's a Saturday, so be sure to mark it in your calendars.

Solar Eclipses

Mercury and Venus are not the only things to pass in front of the Sun. Much more famously, the Moon passes in front of the Sun during a solar eclipse. Just as the Earth orbits the Sun, the Moon orbits the Earth. The planes of the orbit are close to but not exactly the same as each other. When their orbits align, which happens about twice a year, we enter eclipse season. When the three celestial bodies are ordered Sun -> Earth -> Moon, we get a lunar eclipse, and a solar eclipse when they are ordered Sun -> Moon -> Earth. As the astronomy joke goes, there may be a third scenario too, where the objects align as Moon -> Sun -> Earth. Together these three scenarios make up 'lunar eclipse, solar eclipse, apocalypse'. (Disclaimer: this third scenario is not physically possible).

A total solar eclipse is one of the most beautiful things in nature. Even if you have no appreciation of astronomy, the sight of the Sun's corona as the Moon fully blocks the Sun's surface is truly a sight to behold. Total solar eclipses happen only once every 18 months on average and are visible over a very small geographical region about 100 miles across. A solar eclipse occurs during a new moon, which is first noticeable in front of the Sun an hour or two before totality. Slowly, the Moon covers a larger percentage of the Moon, until around 80% of the Sun is covered. Before this percentage is reached, the pupil in the human eye slowly expands to adjust for the lowering light conditions, so you wouldn't notice any decrease in light at all. During this phase, projections of the Sun through trees would also cast crescent shapes on the ground. This part of the eclipse can be viewed safely with eclipse glasses or a solar telescope. As the Moon creeps over even more of the Sun, you'll notice a drop in light and temperature levels. These stages take a long time, but the next phase happens almost in an instant. Over a few seconds, the last sliver of the Sun's surface is blocked by the Moon, revealing for a few seconds the bright red chromosphere through craters on the edge of the Moon. These are called Baily's beads, and last a second or two before they're blocked too and the sky goes dark. Planets and bright stars will then be visible in the daytime sky, and the view of the Sun will appear as you've never imagined it before – no photo can do it justice. The absence of the Sun in the sky gives the appearance of a black velvety hole surrounded by the Sun's glowing corona. Normally, to see the corona, we have to purpose-build a coronagraph to block the Sun from the telescope's view. During a solar eclipse, the Moon is the Earth's coronagraph. This phase of the eclipse is called totality and is perfectly safe to view with your naked eye alone – the only time you can look directly at the Sun. Depending on the specific eclipse and where you are along the eclipse path, totality can last anywhere from 1 to 5 minutes, sometimes even longer. Around you on the ground, animals act strangely during totality. Birds will stop singing, farm animals return to sleep, and frogs and insects start chirping – they all mistake the eclipse for night-time. As totality ends, a brief flash of light is visible as the first beam of sunlight passes through a crater on the edge of the Moon, giving the appearance of a diamond ring in the sky. After this, totality is over, and eclipse glasses must go back on. The partial eclipse then slowly shrinks, before the eclipse ends completely.

Outside the path of totality, a larger geographical region will observe a partial eclipse. Partial eclipses are interesting but lack the striking beauty of a total eclipse. The biggest mistake people make with eclipses is assuming that a 99% partial eclipse is equivalent to a total eclipse – this is not the case. If even 0.01% of the Sun's surface is visible, the sunlight is too bright to witness

Total solar eclipse.

the Sun's corona or any of the other bizarre effects of totality. As was mentioned earlier, total solar eclipses happen somewhere in the world every 18 months, but one is likely to occur in your hometown only once every hundred years, at least. To witness a solar eclipse in your lifetime, as everyone should aim to do, you'll likely have to travel for it. The table below provides a list of all the total solar eclipses up to 2034, with a list of countries where the path of totality passes somewhere through their borders. To see exactly where the eclipse path passes through each country, the website timeanddate.com is an excellent resource.

Total solar eclipses are only possible because the Sun is both 400 times larger than the Moon and 400 times farther away. If the Moon were smaller in the sky, which can happen at some points throughout its orbit, a total solar eclipse could not happen. There are no other planet–moon combinations in the solar system where total solar eclipses occur.

The final category of solar eclipse is the annular eclipse, also known as the 'ring of fire' eclipse. Annular eclipses occur due to variations in the Moon's orbit. The Moon is not always the same size in the sky, because its orbit is not perfectly circular. A full Moon at its closest approach to Earth is a supermoon, about 7% larger than a micromoon – a full Moon at its farthest approach. If the Earth, Sun, and Moon line up perfectly, but the Moon is at its farthest point from Earth in its orbit, the Moon is not

Annular eclipse.

Total Solar Eclipses, 2023–2034	
Date	**Location**
8 April 2024	Mexico, US, Canada
12 August 2026	Greenland, Iceland, Spain
2 August 2027	Morocco, Spain, Gibraltar, Algeria, Tunisia, Libya, Egypt, Saudi Arabia, Yemen, Somalia
22 July 2028	Australia, New Zealand
25 November 2030	Namibia, Botswana, South Africa, Lesotho, Australia
14 November 2031	Panama
30 March 2033	Russia, US (Alaska)
20 March 2034	Nigeria, Chad, Sudan, Egypt, Saudi Arabia, Kuwait, Iraq, Afghanistan, Pakistan, India, China

Stages of a total solar eclipse.

large enough to block the Sun completely. As this happens, the outer edge of the Sun is visible around the Moon, creating the ring of fire. Total solar eclipses and annular eclipses occur with a similar frequency, about once every 1–2 years, somewhere in the world. Annular eclipses, like partial eclipses, must be observed with safety equipment such as eclipse glasses, solar telescopes, or a pinhole camera.

Lunar Eclipses

Lunar eclipses happen during a full Moon and, because eclipse seasons last for a few weeks, will precede or follow the solar eclipse (which happens at new Moon) by two weeks. A lunar eclipse occurs as the full Moon passes into the Earth's shadow, disappearing partly from view before glowing a faint red. If it weren't for the Earth's atmosphere, the Moon would disappear fully during a lunar eclipse. However, a small fraction of sunlight is refracted around the Earth by the Earth's atmosphere, illuminating the Moon with red light. The red light is the same red light we see during sunrises and sunsets. In fact, if you were standing on the Moon during a lunar eclipse, you'd be watching every sunset and sunrise around the Earth simultaneously as a red ring around the Earth.

Total Lunar Eclipses, 2023–2034	
Date	**Location**
13–14 March 2025	North America, western South America
7–8 September 2025	Eastern Africa, Asia, western Oceania
2–3 March 2026	North eastern Asia, eastern Australia, Pacific, western North America
31 December–1 January 2028/29	North eastern Europe, Asia, Oceania, northern North America
25–26 June 2029	Eastern North America, South America, Atlantic, western Africa
20–21 December 2029	Northern North America, Europe, Africa, western Asia
25–26 April 2032	Eastern Asia, Oceania
18–19 October 2032	Eastern Europe, eastern Africa, Asia, north western Oceania
14–15 April 2033	Eastern Africa, western central Asia, western Oceania
7–8 October 2033	North eastern Asia, eastern Oceania, Pacific, western North America

Lunar eclipses take place with a similar frequency to that of solar eclipses but are visible over a far larger area and for far longer durations. Lunar eclipses can take hours between start and finish. The chances are there will be a lunar eclipse near you within the next decade, if you keep an eye out for it. You don't need any special equipment to view a lunar eclipse, simply bring yourself, appropriate clothing for the location and time of year, and a sense of wonder. Gear like cameras, binoculars or telescopes will enhance your experience, just as they would in observing the regular, non-eclipsed Moon, but are not essential.

The table on the page opposite outlines the list of lunar eclipses up until 2034. Lunar eclipses are widely visible, so this table provides only the general continental areas that the event will be visible from. For full maps and

Lunar eclipse.

Sungrazing comet imaged by SOHO/LASCO.

locations, check out the eclipse maps at timeanddate.com.

Solar and lunar eclipses have their own cultural significance, historical dramas, and contribution to modern science.

Sungrazing Comets

The Sun, planets and moons are not alone in the solar system. They are accompanied by an enormous number of smaller bodies also orbiting the Sun. To date, we know of six dwarf planets, nearly 4,000 comets, and 1 million asteroids also orbiting the Sun. The true number of asteroids and comets is far larger, and we expect there to be billions of these smaller objects with an orbit out beyond Neptune. There are three main reservoirs of these small bodies in the solar system. The asteroid belt, composed mainly of – you guessed it – asteroids, sits in an orbit between Mars and Jupiter. The Kuiper belt, another belt of objects in a plane close to the planets, lies beyond Neptune. And finally, the Oort cloud. Unlike the other two regions, the Oort cloud does not exist in the same plane as the planets, but rather in a sphere containing rocky and icy objects orbiting between 2,000 and 200,000 times farther away from the Sun than the Earth. The Oort cloud is so large that objects within it may even interact with Oort clouds of other stars. When new comets are discovered, they are often determined to have originated from the Oort cloud, having taken hundreds of thousands of years to reach the inner solar system. There are some famous comets with shorter periods, like Halley's Comet, which famously visits the inner solar system every 75 years or so, and is next visible in the sky in July 2061. More recently, Comet Neowise graced our skies in summer 2020, a beacon of wonder in otherwise uncertain times. Comets are solid, icy bodies for most of their orbit. It is as they approach the inner solar system and the heat of the Sun that some of their icy surface vaporises to form the comet's tail. (Comets have two tails, one blown away radially from the Sun and another electrically charged tail following the magnetic field of the solar

A sundog

wind.) When coronagraph telescopes like LASCO observe the Sun, occasionally new or unknown comets will enter the field of view of the telescope. During this time, they glow bright before we lose sight of them in front of or behind the Sun. We call these comets sungrazing comets. They are unpredictable, as their initial discovery is usually by LASCO – which means that they weren't known before. Occasionally, a comet will survive this encounter as we witness it emerge from the other side of the Sun to live for another orbit. But more often than not, their disappearance behind the Sun from our perspective is the last time they're ever spotted, as they evaporate in their entirety on their closest approach to the Sun. From that moment on, there is one less comet in the solar system.

Atmospheric Effects

Sometimes, the beauty of sunlight surprises us when we least expect it. It is often easy to forget that the Sun described in the pages of this book with its nuclear fusion and solar flares is the same Sun sitting above us on a sunny day. Although we cannot see these phenomena with our eyes, the intricacy of sunlight is all around us. The sky is blue because of the scattering of shorter wavelengths of sunlight in the atmosphere and red at sunset because more air particles sit between us and the Sun. On a rainy day, the white light of the Sun splits up into its colour components, projecting a spectrum of light in the sky – a rainbow. The splitting of sunlight during a rainbow is the same physics used in telescopes to study light from the Sun. Under specific conditions, with ice crystals in the upper atmosphere, sunlight can also produce spectra around the Sun, in a rare meteorological effect called a sundog (also known as a 'mock Sun' or parhelion).

Next time you're outside on a sunny day and feel the warmth of sunlight hitting your face, remember – the light waves heating your skin are the same light waves used by astronomers to understand further the mysteries of our local star.

6. RESOURCES and GLOSSARY

Online Resources and Software

- **Helioviewer - www.helioviewer.org**
  The online version of the below software, with reduced capabilities

- **JHelioViewer - www.jhelioviewer.org**
  Software to make videos and images of the Sun, from various space-based telescopes

- **Solar Monitor - www.solarmonitor.org**
  Website showing up-to-date images from the Sun

- **The Sun Now (SDO) - sdo.gsfc.nasa.gov/data/**
  Most recent images from the NASA Solar Dynamics Observatory

- **Time and Data - www.timeanddate.com/eclipse/**
  List and locations of all upcoming eclipses

- **American Astronomical Society - www.eclipse.aas.org**
  Resources for observing solar eclipses safely

- **www.swpc.noaa.gov/products/goes-x-ray-flux**
  Live soft X-ray measurements of the Sun, showing recent solar flares

- **Met Office – https://www.metoffice.gov.uk/weather/learn-about/space-weather/uk-scales**
  Full tables providing information on space weather impacts

- **Met Office – https://www.metoffice.gov.uk/weather/specialist-forecasts/space-weather**
  Bi-daily space weather forecast

Glossary of Terms

Absorption spectrum	A spectrum with a pattern of dark lines produced by the absorption of photons by plasma primarily in the photosphere and chromosphere.
Active region	A region of the corona with a strong and complex magnetic field, typically found above a sunspot region.
Atom	The building blocks of our universe, consisting of a nucleus (protons and neutrons), surrounded by a cloud of electrons.
Aurora	An illumination of the Earth's atmosphere upon the impact of high-energy particles, manifesting itself as the northern lights (aurora borealis) and southern lights (aurora australis) in the northern and southern hemisphere respectively.
Barycentre	The centre of gravity between objects of mass which both objects orbit.
Blackbody	A perfect emitter and absorber of radiation, with a peak wavelength dependent on the object's temperature.
Black hole	The end state of the largest stars, with an infintely dense point at the centre (a singularity). They are so dense, not even light can escape. The boundary of a black hole is called the event horizon.
Bow shock	A shock originating from an abrubt discontinuity in density, pressure, and temperature found ahead of the magnetopause and heliopause.
Carrington event, the	The largest geomagnetic storm on human record, on September 1st–2nd 1859. It is considered the modern worst-case scenario for space weather forecasters.
Chromosphere	The base of the Sun's atmosphere, with the lowest temperatures on the Sun.
Corona	The main region of the Sun's atmosphere, with temperatures around 1 million degrees.
Coronal heating problem	The unsolved mystery of why the Sun's corona is hotter than the photosphere.
Coronal hole	A region of the corona with an open magnetic field allowing fast solar wind to emanate.

Coronal mass ejection (CME) An eruption of plasma from the Sun's atmosphere.

Dipole The most basic magnetic field, with one north pole and one south pole.

Doppler shift The apparent shift in wavelength between a wave source and an observer, dependent on the relative velocity between them. Both light and sound can be Doppler shifted.

Dungey cycle The cycle of magnetic reconnection between the solar wind and Earth's magnetosphere, which leads to the formation of the aurora.

Electromagnetic spectrum The different wavelengths of light from gamma rays to radio waves.

Electromagnetism A force encompassing electric and magnetic forces occuring between objects with charge that is one of the four fundamental forces.

Electron degeneracy The force between electrons resisting the collapse of a white dwarf star.

Emission spectrum Spikes in the Sun's spectrum produced by the emission of photons from plasma primarily in the corona.

Event horizon The outer boundary of a black hole, defining the region from which light cannot escape.

Exoplanet A planet orbiting a star other than our own Sun.

Filament Dense plasma suspended in the solar atmosphere and visible against the solar disc (a prominence viewed from a different angle).

Flux rope A complicated and twisted magnetic structure in the solar atmosphere.

Geoeffectiveness The ability of the solar wind or a CME to cause a geomagnetic storm on Earth.

Geomagnetic storm A disruption of the Earth's magnetic field due to the impact of a CME or fast solar wind, which causes the aurora and disruption to technology.

Goldilocks zone The region around a star where temperatures are potentially suitable for liquid water to exist.

Granulation Fine structure in the photosphere created by convection beneath the Sun's surface.

Gravity	A force providing attraction between objects with mass that is one of the four fundamental forces.
Halo CME	A CME viewed head-on, heading either directly towards, or away from us.
Heliopause	The boundary between the front (in direction of motion) of the heliosphere and interstellar space.
Heliosphere	The region encompassed within the Sun's magnetic field and containing the Sun and all of the planets.
Hertzsprung–Russell diagram	A diagram showing the brightness and temperature of different varieties of stars.
Ion	An atom with one or more electrons stripped away, which gives it a positive charge.
Lunar eclipse	A phenomenon caused by the full Moon passing into the Earth's shadow, which causes refracted red light to illuminate the Moon.
Magnetic reconnection	The reconfiguring of a magnetic field to a lower energy, which releases energy in the process.
Magnetohydrodynamics	The study of plasma under the influence of both fluid mechanisms and electromagnetism.
Magnetopause	The boundary between the Sun-side of the magnetosphere and the solar wind.
Magnetosphere	The magnetic field surrounding a planet, such as the Earth.
Magnetotail	The night side of the magnetosphere stretched out by the solar wind.
Main sequence star	A star in the main stage of its lifetime, which is defined by the presence of hydrogen to helium nuclear fusion in the core.
Maunder minimum	A prolonged period of low solar activity between 1645 and 1715.
Nanoflare	A very small solar flare, far below the level of detection possible with current instrumentation - a potential heating mechanism for the solar corona.
Nebula	A gaseous cloud in space that has a fuzzy appearance when viewed through a small telescope.

Neutron degeneracy The force between neutrons resisting the collapse of a neutron star.

Neutron star The end state of medium to large stars, whose structure is maintained by neutron degeneracy.

Nuclear fission The process of splitting two heavier elements into a lighter one, producing energy for elements heavier than iron.

Nuclear fusion The process of combining two lighter elements into a heavier one, producing energy for elements lighter than iron.

Ozone layer A layer of the stratosphere responsible for blocking the majority of harmful UV radiation from the Sun.

Parker spiral The spiral structure within the solar wind caused by the rotation of the Sun at the base of the solar wind.

Penumbra The light edge around a sunspot.

Perihelion The closest approach of an object's orbit to the Sun.

Photon A packet of energy travelling in a light wave.

Photosphere The surface of the Sun producing the sunlight we see.

Plage A slightly lighter patch around a sunspot.

Planetary nebula A nebula created by the expulsion of the outer layers of a small to medium star at the end of its life.

Plasma A fluid of high-temperature electrons and ions with a neutral charge on large scales.

Pore A small, sunspot-like region with no penumbra.

Prominence Dense plasma suspended in the solar atmosphere and visible over the solar limb (a filament viewed from a different angle).

Quarks An elementary particle, which when combined together, create protons and neutrons.

Radio blackout The failure of radio transmission due to the Earth's upper atmosphere expanding under the impact of incident solar flare X-rays.

Red giant A small to medium star in the stage of life in which it expands after beginning the nuclear fusion of helium into the next few heavier elements.

Red supergiant A large star in the stage of life in which it expands during nuclear fusion of all heavier elements up to iron.

Ring fusion The reignition of the nuclear fusion of hyrdrogen in a ring around a star's core, following the star's slight collapse as its core hydrogen is depleted.

Seeing The effect of the air above us on the quality of astronomical observations.

Single event upset A technical error created by the collision of a rogue electron, such as from a solar flare, with an electronic system.

Solar continuum The Sun's blackbody spectrum, depending on its temperature, before any spikes/dips in the spectrum are created due to absorption/emission of atmospheric plasma.

Solar cycle An approximately 11-year cycle of increasing and decreasing solar activity.

Solar disc The main circle of the Sun that we observe.

Solar dynamo The mechanism within the Sun creating its global magnetic field.

Solar eclipse A phenomenon caused by the new Moon passing between the Earth and Sun. Totality occurs when the Sun is fully covered by the Moon, with no solar surface showing.

Solar energetic particles Protons and electrons accelerated to high speeds and energies during a solar flare, capable of interferring with electronic systems.

Solar flare An instance of plasma heating, particle acceleration, and light emission in the solar atmosphere as a result of magnetic reconnection.

Solar limb The edge of the Sun, from our pespective.

Solar radiation storm The impact of solar energetic particles on Earth that causes disruption to electronic systyems.

Solar wind A constant stream of plasma emanating from the Sun's atmosphere, travelling between 400-700 km/s.

Space weather The effect of the Sun's behaviour on the Earth and near-Earth environment.

Spectropolarimetry The study of polarisation in the Sun's spectra, providing insight into the Sun's magnetic fields.

Stealth CME A CME with no apparent source region on the solar disc.

Stellar flare A solar flare taking place on a star other than our own Sun.

Strong nuclear force A force holding quarks together to make up protons and neutrons that is one of the four fundamental forces.

Sunquake Vibrations on the Sun's surface created by the explosive release of energy during solar flares.

Sunspot A cool, dark region in the Sun's photosphere caused by a strong magnetic field preventing the inflow of hot plasma.

Superflare The very largest stellar flares, hundreds to thousands of times stronger than the largest solar flares.

Supernova A large explosion caused by the death of a large star as it attempts the nuclear fusion of iron, which cannot provide the energy needed to resist gravitational collapse.

Transition region The thin region between the chromosphere and corona, where temperatures increase rapidly.

UK National Risk Register A list of all potential disasters in the United Kingdom ranked by impact and probability of occurence.

Umbra The dark patch in the middle of a sunspot.

Wave-heating The conversion of energy from vibrations along a magnetic field to plasma heating - a potential heating mechanism in the solar corona.

Weak nuclear force A force holding protons and neutrons together in the nucleus of an atom that is one of the four fundamental forces.

White dwarf star A small to medium star in its end state, in which its structure is maintained by electron degeneracy.

Zeeman splitting The splitting of a spectral line in the presence of strong magnetic fields.

Acknowledgments

5	© Ryan French
6	Wikimedia Commons/Matthew Field
7 ***top***	Wikimedia Commons
7 ***bottom***	Wikimedia Commons/British Museum
8	Wikimedia Commons
9 ***top***	Wikimedia Commons/John of Worcester
9 ***bottom***	The Galileo Project/M. Kornmesser
10–11	Wikimedia Commons
12	Wellcome Collection
13	NASA/JPL/Space Science Institute
15	© Ryan French
17 ***top***	© Ryan French
17 ***bottom***	Wikimedia Commons/Richard Carrington
19	© Vincent Ledvina
20 ***left***	Wikimedia Commons
20 ***right***	NASA
21	Edward Walter Maunder/Annie Russell Maunder
22	NASA
23 ***top***	Wikimedia Commons
23 ***bottom left***	Wikimedia Commons
23 ***bottom right***	© Ryan French
25, 27	© Ryan French
28	Wikimedia Commons/Igor da Bari
29	NASA/ESA/CSA/STScl
31	EHT Collaboration
32	© Ryan French
34 ***top left***	Wikimedia Commons/Brezhnev30
34 ***top right***	NASA/ESA/JPL/Arizona State Univ.

34 ***middle left***	Shutterstock/Lucas Gojda
34 ***middle right***	NASA
34 ***bottom left***	Wikimedia Commons/User:Colin
34 ***bottom right***	NASA Goddard
36	Wikimedia Commons/Wing-Chi Poon
37	© Ryan French
39	N.A.Sharp, NOAO/NSO/Kitt Peak FTS/AURA/NSF
40–41	© Ryan French
42	Shutterstock/Awe Inspiring Images
43–44	NASA Solar Dynamics Observatory
45	© Ryan French
47–48	© Ryan French
50	NASA
51 ***top left***	NASA/JPL
51 ***top right***	NASA/JPL-Caltech
51 ***middle left***	NASA/JPL
51 ***bottom left***	NASA/JPL
52	NASA/JPL-Caltech
53 ***top***	STEREO/NASA
53 ***bottom***	NASA/Johns Hopkins APL/Naval Research Laboratory
54	© Ryan French
55	NASA Goddard
56 ***top***	© Vincent Ledvina
56 ***bottom***	© David Wildgoose
57	NASA/GSFC/Solar Dynamics Observatory
58	NASA Goddard
63 ***left***	NASA's Goddard Space Flight Center/Ludovic Brucker
63 ***right***	Wikimedia Commons/Arnoldius
64 ***top***	ESO/K. Ohnaka

64 *bottom*	ESO/M. Montargès et al.
65	NASA/JPL-Caltech
67	NASA Goddard
68	© Ryan French
69 *top*	© Ryan French
69 *bottom*	Solar Orbiter/EUI Team/ESA & NASA; CSL, IAS, MPS, PMOD/WRC, ROB, UCL/MSSL
70 *top*	Solar Orbiter/EUI Team/ESA & NASA
70 *bottom*	© Ryan French
71, 73	NSO/AURA/NSF
76 *top left*	Wikimedia Commons
76 *top right*	Wikimedia Commons/Julius Söhn
76 *bottom*	Wikimedia Commons/Anoop K R
77–78	© Tom Kerss
79	Wikimedia Commons/Ben P L
81–82	© Ryan French
84	NASA Goddard
85	NASA/GSFC/Solar Dynamics Observatory
86 *top*	Wikimedia Commons/BBSO/NJIT

86 *left* NASA/GSFC/Solar Dynamics Observatory

86 *right* NASA Goddard

87 *top left* NASA Goddard

87 *top right* NASA Goddard

87 *bottom left* NASA Goddard

87 *bottom right* NASA/GSFC/Solar Dynamics Observatory

88 © Ryan French

89 © Ryan French

90–91 NASA Goddard

92 NASA/Aubrey Gemignani

93 Wikimedia Commons/A013231

94 NASA/Aubrey Gemignani

95 © Ryan French

96 NASA Goddard

97 Wikimedia Commons

Specialist editorial support was provided by Dr Ed Bloomer, Senior Astronomy Manager, Digital & Data at Royal Museums Greenwich. Editorial support also provided by Louise Jarrold, Publishing Executive at Royal Museums Greenwich.

Collins titles by Tom Kerss

Stargazing | 9780008196271

Offering complete advice from the ground up, Stargazing is the perfect manual for beginners to astronomy – introducing the world of telescopes, planets, stars, dark skies and celestial maps.

Discover how to tackle light pollution, how to stargaze with just your eyes, and what equipment is best for beginners. Explains the best ways to plan your stargazing experience and the key things to look out for on specific dates throughout the year.

Moongazing | 9780008305000

An in-depth guide for all aspiring astronomers and moon observers with detailed moon maps. This book covers the history of lunar exploration, the properties of the moon, its origin and orbit. This is the ideal book for Moon observers covering essential equipment, and the key events to look out for.

A comprehensive section covers astrophotography using lenses, telescopes, Smartphones, including video and how to process your images. Comes with a photographic atlas of lunar features with plates and annotated maps.

Look to the skies with Collins

Planisphere | 9780007540754

Easy-to-use practical tool to help astronomers identify the constellations and stars every day of the year. The *Collins Planisphere* includes information on planetary positions until 2028, as well as suggestions for objects to view at different times of the year.

Northern Lights | 9780008465551

Discover the incomparable beauty of the Northern Lights with this accessible guide for aspiring astronomers and seasoned night sky observers. This book covers the essential equipment needed for observation and photography and is full of stunning photographs.

Learn about the formation, properties and types of auroras, the mythology and history of the northern lights, the evolution of aurora science from antiquity to the modern day, and how to capture your own images.

Northern Lights will give you a complete understanding of one of the true wonders of the natural world and leave you prepared for the adventure of a lifetime.

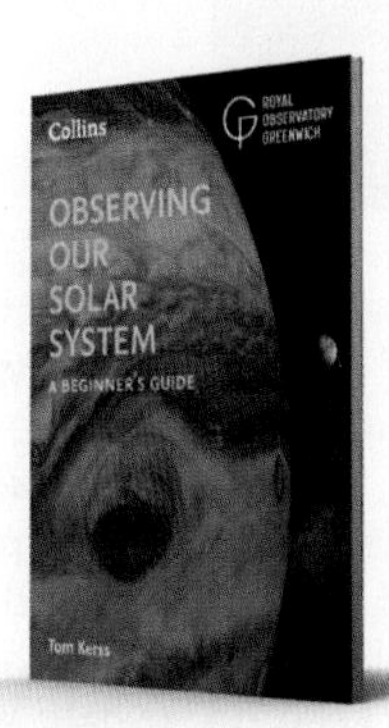

Observing Our Solar System | 9780008532611

Study the ever-changing face of the Moon, watch the steady march of the planets against the stars, and witness the thrill of a meteor shower, or the memory of a once-in-a-generation comet.

Find out how the Solar System came to be understood, from ancient theories to modern times, discover how to explore the Moon, Sun and planets by eye and telescope, and learn about the more obscure elements of the night sky, from asteroids to conjunctions to eclipses.

Officially approved by the Royal Observatory Greenwich, *Observing Our Solar System* is the perfect gift for all amateur and seasoned astronomers.

AVAILABLE TO BUY FROM ALL GOOD BOOKSELLERS AND ONLINE

facebook.com/CollinsAstronomy

Author Biography

Dr Ryan French is a solar physicist, science communicator, and author. He is pursuing the mysteries of the Sun at the forefront of modern solar physics research, using cutting-edge telescopes on the ground and in space. His research takes him all over the world, collaborating with the global community of solar physicists. Ryan also works to share the wonders of the Sun and space with the public through social media, museums, and observatories and on television and radio. He is also an avid hiker, rock climber, and skier, perhaps because the mountains take him closer to the Sun. Keep up to date with Ryan's other projects at www.ryanjfrench.com